The
Endangered
Species Act

History, Implementation, Successes, and Controversies

The Endangered Species Act

History, Implementation, Successes, and Controversies

J. Peyton Doub

CRC Press
Taylor & Francis Group
Boca Raton   London   New York

CRC Press is an imprint of the
Taylor & Francis Group, an **informa** business

CRC Press
Taylor & Francis Group
6000 Broken Sound Parkway NW, Suite 300
Boca Raton, FL 33487-2742

Version Date: 20120817

International Standard Book Number: 978-1-1383-7467-6 (Paperback)
International Standard Book Number: 978-1-4665-0737-1 (Hardback)

Library of Congress Cataloging-in-Publication Data

Doub, J. Peyton.
   The Endangered Species Act : history, implementation, successes, and controversies / J. Peyton Doub.
     p. cm.
   Includes bibliographical references and index.
   ISBN 978-1-4665-0737-1
   1. United States. Endangered Species Act of 1973. 2. Endangered species--Law and legislation--United States. I. Title.

   KF5640.D68 2013
   346.7304'69522--dc23
   2012016412

Visit the Taylor & Francis Web site at
http://www.taylorandfrancis.com

and the CRC Press Web site at
http://www.crcpress.com

To my parents, William and Mary Graham Doub, who taught

me to appreciate the natural world as a gift from God.

Contents

Preface

Protecting animal and plant species from extinction is one of the most visceral challenges of the modern environmental movement. Issues such as climate change and waste management are no less important, and indeed they themselves are highly integral to species protection, but in many ways these are academic issues that lie mainly in the realm of professionals and professors. But, even most nonprofessionals, not conversant in the vocabulary and principles of the environmental profession, delight in the diversity of wildlife and wildflowers that surround them and care if that diversity may one day be unavailable to them. While so caring, they may also express concern that protecting species comes at a price, perhaps to the value of their own property or at least to the interests of their employers or the business interests in which they invest. The subject of this book, the federal Endangered Species Act, elicits a diversity of passionate responses from professional and average citizen alike but rarely does it elicit indifference.

Ask many average people on the street about the National Environmental Policy Act (NEPA) or Resource Conservation and Recovery Act (RCRA) and you are likely to encounter a lot of blank stares, but ask about the Endangered Species Act and you will almost assuredly spark a debate. Most Americans value iconic species such as the bald eagle, a species that may, arguably, owe its continued existence and indeed its present abundance to the Endangered Species Act. In 2005, the government declared the bald eagle recovery complete enough to remove (delist) it from the group of species directly protected by the Endangered Species Act. And most American bird-watchers, a hobby of understated importance to more of the American population than ever suggested by the media, place similar value on the whooping crane—a species that almost undoubtedly owes its continued existence to the Endangered Species Act but, unlike for the bald eagle, a species whose recovery story remains substantially unwritten and uncertain. From a depth of fewer than 40 remaining birds in the 1970s, its population attained levels over 400, only to inexplicably slide back below 300. Some of the more dedicated bird-watchers even hold out hope that the ivory-billed woodpecker, a species considered extinct by most authorities for several decades, might still be clinging to existence in some of the remoter swamps of the American southeast. If so, it also might one day owe its existence to protections afforded to it and its habitat under the Endangered Species Act.

But then, there are certain loggers, and former loggers, who—rightly or wrongly—blame the demise or eventual demise of their vocations to protection of old-growth forests directed by the Endangered Species Act to conserve yet another bird species, the northern spotted owl. Many of these individuals are not interested in arcane rationales for preserving some obscure bird they cannot even recognize or, worse, arguments that they would eventually lose their jobs anyway once no more old-growth forest remains. Then, there are certain ranchers who are unwilling to accept the loss of a portion of their product, whether sheep or cattle—the fruit of their labor—to protect the gray wolf. And while most real estate developers have long come to grudging acceptance that a portion of their inventory—land—must be set aside, actually paid much like a tax, to help protect wetlands, floodplains, and in some areas even scenic vistas, even they can reach their boiling point when some regulator unexpectedly pops yet another additional land set-aside because they had the misfortune of buying property used by some obscure frog, plant, or insect.

This book delves into the intricacies of the Endangered Species Act—its history, requirements, controversies, popular elements, and unpopular elements—and is written to an audience of professional environmental practitioners and interested members of the public alike. The book is written from the perspective of an environmental practitioner who must work with the Endangered Species Act in the course of performing routine professional duties. The book is not a regulatory guidance document as might be written by a regulatory official with the U.S. Fish and Wildlife Service or other agency with direct regulatory authority under the Endangered Species Act. The book is neither an academic nor a legal treatise; it is written by neither a researcher nor a lawyer. The book is also not a discourse on policy direction or current events involving the Endangered Species Act, even though portions of the book do deal with current and historical policy issues. Regulatory analysts, biological researchers, environmental lawyers, and environmental policy analysts may, however, find the book interesting, especially with respect to how their professional activities fit into the Endangered Species Act compliance process as practiced today.

It is hoped both supporters of the Endangered Species Act (who passionately feel that everything that walks, flies, crawls, or flowers should be carefully preserved) and antagonists of the Endangered Species Act (hardworking people who likely value the trees and birds around their homes but just want to be able to pursue their vocations and provide for their families without the government getting in their way) will find the book interesting and informative. It is a book about a scientific subject and hence uses scientific terminology, both for engaging the professional and for educating the interested layperson. But, it is written to be accessible to both, and it is hoped anyone interested in the Endangered Species Act and the issues of species protection and biodiversity should find the book understandable.

The book relies on the experience of an environmental professional to outline both the basics of the act and some of the more interesting stories and controversies of the act. Like other environmental regulations, the Endangered Species Act offers protections that much of the public desires, but it does so at a price—a price measured not only in taxpayer and industry dollars but also in costs and inconveniences to the public, to a public that includes property owners and persons employed by, or otherwise invested in, property owners. The story to be told then is not only a story of science but also a human story.

Author

Peyton Doub has more than 20 years' experience as an environmental consultant working with the Endangered Species Act and related environmental regulations and 4 years working on the environmental staff of a federal agency, the Nuclear Regulatory Commission. He is a certified environmental professional (CEP), professional wetland scientist (PWS), and a qualified professional under the Maryland Forest Conservation Act, and his education includes an MS in plant physiology from the University of California at Davis and a BS in plant sciences from Cornell University.

Mr. Doub has performed dozens of biological surveys, such as wetland delineations, forest inventories, and vegetation surveys as a consultant to numerous federal agencies, developers of energy facilities, and housing developers. He has also contributed biological expertise to numerous environmental impact statements (EISs) and environmental assessments (EAs) under the National Environmental Policy Act and remedial investigations and feasibility studies (RI/FSs) under the Comprehensive Environmental Response, Compensation, and Liability Act (CERCLA; better known as Superfund). Further, he has contributed to the design of several restoration plans for restoring wetlands, forests, and other habitats of value to endangered species and other ecologically valuable resources.

1

Roots of Endangered Species Conservation

1.1 Introduction

Most Americans know of the Endangered Species Act, many admire its objectives, and some are aware of some of its successes, especially with respect to once nearly extinct but now relatively common and iconic species such as the bald eagle (*Haliaeetus leucocephalus*), American bison (*Bison bison*), and American alligator (*Alligator mississippiensis*). Some others conversely know of the act in a negative sense—one more government regulation that increases the cost of developing land, building infrastructure, or otherwise carrying out the economic business of our land. Talk to these Americans and the conversation will likely turn to the snail darter (*Percina tanasi*), northern spotted owl (*Strix occidentalis caurina*), or—more recently—the polar bear (*Ursus maritimus*); that is, it will turn to species whose conservation is perceived to stand in the way of "progress." Yet, some other Americans view the act with frustration, pointing to species such as the whooping crane (*Grus americana*) pictured on the cover or California condor (*Gymnogyps californianus*), which despite decades of protection and costly conservation effort under the act have yet to show promising rebounds in their population.

1.2 Purpose and Objectives of This Book

Few Americans, regardless of whether they see themselves as defenders or as victims of the Endangered Species Act, really understand the act, the science that underlies the objectives of the act, or how the act has been implemented in the nearly four decades since its initial promulgation in 1973. Such understanding has long been the province of a cadre of professional biologists, ecologists, environmental scientists, and environmental lawyers who work for government agencies, consulting firms, and industry in developing, administering, and complying with the act and its associated bevy of regulations. It is hoped this book will help students aspiring to these fields; beginners starting out in these fields; landowners and business owners affected by the act; nature and conservation enthusiasts such as bird-watchers, hunters, and fishers; and interested members of the public to gain a better understanding and appreciation for one of America's premier environmental conservation laws—the one specifically and directly scoped with preventing extinction—the permanent loss of our nation's biodiversity. It should also help provide new insight and perspective on the act to experienced consultants and other scientific practitioners. The book is written from the perspective of an environmental consultant who must assist clients with complying with the act. It is not written from the perspective of those who write, administer, make decisions, or otherwise exert regulatory power under the act; it is instead written from the perspective of one who must know and live with the provisions of the act.

This book is not intended to be a cheerleader for the act or even the objectives of the act. It will of course highlight the noble aspirations of the act and note some of the act's successes. But, it will also note some of the problems inherent in the act, discussing the inevitable conflicts that have arisen and continue to arise as implementation of those noble aspirations butt up against the realities of our economy and our rights to property ownership as guaranteed in our Constitution. Like other environmental regulations, the Endangered Species Act is an imperfect attempt to balance the interests of those who admire the natural world as the shared heritage of us all and those who work to meet the economic challenges of supporting our advanced standard of living. The Endangered Species Act, as presently written and administered, is not the only possible way, and probably not the best way, to balance these objectives. But a comprehensive understanding of the act and how it functions in its present form is necessary to serve as a foundation for proposing improvements to the act. Unless employed by environmental advocacy groups such as Greenpeace or the World Wildlife Fund, most environmental consultants and other technical specialists involved in environmental planning are not outspoken environmental activists with clearly expressed "proenvironment" or "progrowth" opinions. Their job is to provide scientific expertise, not opinions.

1.3 Early Roots of Conservation

When promulgated in 1973, the objectives of the act were not entirely novel. In fact, one may view the Endangered Species Act as simply the modern American version of various laws, decrees, and practices that have existed over the years to ensure the continued availability of plants and animals viewed as desirable by civilization. Kings of England and other European countries established royal forests and deer parks as early as the Middle Ages to prevent commoners from hunting prized game, especially red deer.[1] The king's concern, of course, was not the ecological consequences of fewer or no deer; instead, the concern was the continued availability of plenty of deer for his and his noble's own pursuit. The target of the American Revolution, King George III, issued decrees limiting the colonist's ability to harvest tall trees, especially prized white pine (*Pinus strobus*) trees, needed as masts for royal shipbuilding. The Crown government marked the best white pine trees with the "king's mark," indicating that the tree could be cut for royal use only, not by the local population.[2] Again, the king could care less about protecting white pine as a species; ensuring continued shipbuilding was the only objective.

The newly independent nation had a seeming abundance of natural resources, and far more immediate challenges related to survival and an insatiable drive to expand, than conservation of what it perceived as more of a blank canvas. Only with the disappearance of the "frontier" did any nostalgia for the once-ubiquitous wilderness arise. The earliest roots of American environmental conservation are discussed in the introduction to a new book on environmental assessments authored jointly by Charles Eccleston and myself.[3] One of the earliest American environmental successes involved the 1872 enactment of the law establishing Yellowstone National Park, the world's first national park, thereby setting a precedent for the preservation of scenic federal lands. The very concept of a park ultimately reflects the "parks" or "enclosures" of naturally vegetated lands in Europe established for hunting by nobility. In 1873, the American Association for the Advancement of Science petitioned Congress to halt the unwise use of natural resources. In the following years, Congress continued to lay the foundation for federal protection of lands by expanding the national park system, establishing national forests and the U.S. Soil Survey. Many states followed suit with the establishment of state parks and state forests. In 1891, John Muir founded the Sierra Club, one of the founding organizations of the environmental movement, which has remained active ever since. While the setting aside of lands in the late nineteenth century for conservation purposes constituted a major advance toward protecting rare and declining species from extinction, the nation was still far from the notion of enacting regulations such as the Endangered Species Act that are directed at protecting individual species from harm and exploitation.

Public concern driving much of the earliest attempts to conserve species and natural lands in the late nineteenth and early twentieth centuries was driven to a large degree by the rapid and stunning declines in what had once been some of the most common bird species. Chapter 10 of the Cornell Lab of Ornithology *Handbook of Bird Biology* provides a good discussion of how bird population declines underpin early public concern that leads to American conservation laws.[4] Around the turn of the twentieth century, Americans began to notice, and seemed quite puzzled by, the disappearance of what was once an almost unimaginably abundant species: the passenger pigeon (*Ectopistes migratorius*). To one with a modern perspective, imagining the extinction of the passenger pigeon would be like imagining the extinction of corn, crabgrass, or house flies—how could we ever run out, even if we tried? Southeastern farmers felt that they could only dream of the extinction of the Carolina parakeet (*Conuropsis carolinensis*), a colorful attractive bird that had an undesirable habit of feeding on grain—until by the early twentieth century they realized they got their dream. The Carolina parakeet was a pest; a modern perspective on its possible extinction would be like imagining the extinction of the gypsy moth (*Lymantria dispar*). But, nonfarmers who enjoyed seeing this beautiful bird without experiencing the economic losses of its feeding habits began to miss its presence.

The concept of setting aside undeveloped land for the sole purpose of protecting a species of no direct economic value was anathema to many, if not most, Americans in the nineteenth century. The establishment of Yellowstone National Park in 1872 was the start of a slow inflection from a focus on national resource exploitation to wise natural resource management and conservation, but protection of wilderness aesthetics was the primary purpose for establishing the park, not protecting wildlife and other plant and animal species of no use for agriculture or hunting. An even more watershed incident was the setting aside by Theodore Roosevelt in 1903 of Pelican Island for the sole purpose of protecting the brown pelican from hunters (*Pelecanus occidentalis*). Pelican Island, now Pelican Island National Wildlife Refuge, was the start of the National Wildlife Refuge System, which protects large areas of important natural habitats for species protected under the Endangered Species Act. Interestingly, the brown pelican, which had been listed under, and hence protected under, the Endangered Species Act, in 2009 was determined to be recovered and hence delisted (removed from the protections of) the act.[5] The brown pelican, whose populations had once been severely reduced due to hunting and use of the pesticide DDT, is now a common sight along the coasts of the southeastern states. The agricultural insecticide DDT, widely used in the United States from the 1940s to the 1970s, is recognized as having contributed to population losses of many predatory bird species, including the bald eagle. The brown pelican is a highly visible sign of the successes of American conservation efforts, including the Endangered Species Act.

1.4 History of American Conservation and Endangered Species Legislation

The following brief summary of American conservation and endangered species legislation serves to introduce how the Endangered Species Act evolved. Many of the regulatory efforts introduced are discussed further in

Chapter 4 of this book, concerning the interrelations between the Endangered Species Act and other environmental protection requirements that resource managers must deal with in the United States.

Perhaps the earliest of the earliest environmental protection acts that still remain in effect today is the Rivers and Harbors Act of 1899. In its present form, the act is codified in 33 U.S.C. § 407. This simple act was never intended to be a conservation measure. It only established a requirement that anyone performing work in the navigable waters of the United States first obtain a permit from the U.S. Army Corps of Engineers. The intent was purely directed at preserving navigation; preserving natural habitats, natural resources, or biodiversity could not have been further from the minds of the act's framers. Still, the act foreshadowed perhaps one of the most far-reaching and controversial environmental conservation acts of the second half of the twentieth century: the Clean Water Act. Enacted in 1972, the Clean Water Act included provisions requiring permits from the Corps of Engineers for filling in any waters of the United States. Subsequent court decisions interpreted these provisions as encompassing most work in most surface water features, whether navigable or not, including wetlands adjacent to those waters. Many species listed as threatened or endangered under the Endangered Species Act (see Chapter 3 of this book for a detailed discussion of how the terms threatened and endangered are defined under the act) require or prefer wetland habitats, and the requirement for a federal permit triggers provisions of the Endangered Species Act requiring interagency consultation.

The visible and rapid decline of the passenger pigeon and many other game bird species in the late nineteenth century prompted passage of the Lacey Act of 1900.[6] The Lacey Act prohibited interstate (or interterritorial) transport of birds and other wildlife if in violation of state or territorial law. It sought to head off the depletion of game species valuable to hunting and other forms of commerce. Unlike the modern Endangered Species Act, the Lacey Act as passed in 1900 was driven almost completely by economics, not ecology. It obviously was intended to help prevent extinction of species, but only those valuable to commerce and recreation. Like many early laws of the United States, it largely sought only to ensure that interstate commerce conducted in the federal arena complied with state (or territorial) laws; it did not establish prohibitions that extended beyond those already established by states and did not seek to extend the prohibitions of one state to other states that had not established similar prohibitions.

Some of the framers of the Lacey Act might have wanted to establish some uniform federal standards on allowable killing of game species that could be enforced across the entire country, but they would have undoubtedly viewed this goal as unachievable. Imposing federal limitations exceeding state requirements was highly controversial at the time because the Tenth Amendment of the Constitution states:

> The powers not delegated to the United States by the Constitution, nor prohibited by it to the States, are reserved to the States respectively, or to the people.

The constitutionality of the Rivers and Harbors Act was not highly controversial because it sought to protect the ability of citizens to conduct interstate commerce by ensuring access to navigable waterways. A host of federal laws, including many environmental laws (including but not limited to the Endangered Species Act), passed subsequent to World War II do substantially impose prohibitions and requirements not clearly interrelated with interstate commerce, and the constitutionality of many of those laws remains a vibrant controversy. Even though many modern environmental laws do indeed establish nationwide federal limitations that greatly exceed the corresponding limitations of many of the affected states, they frequently try to tie those limitations to a very broadly defined scope of interstate commerce.

Although over a century old, frequently amended, extensively overhauled in 1981, and partially overlapped by newer and more restrictive statutes, including the Endangered Species Act, the Lacey Act remains in place today among our federal environmental laws. In its present form, the provisions of the Lacey Act are codified in 16 U.S.C. §§ 3371–3378. Its original scope remains largely intact but has been expanded over the decades to encompass prohibitions on wildlife killed in violation of foreign as well as state laws, has been extended to plants as well as wildlife, and perhaps most important has been broadened to prohibit importation of wildlife and plant species that might threaten crop production and horticulture in the United States. Although still driven primarily by economic concerns, the Lacey Act now has an ecological element as well, as invasive species from foreign countries can threaten ecosystems as well as agriculture. If you have ever had to answer questions from customs agents on whether you might be carrying harmful plants or animals, then you were experiencing (at least in part) the Lacey Act.

The visible disappearance of birds in the decades around the turn of the twentieth century, especially wading birds hunted to provide showy feathers, led to another quite different conservation statute, the Weeks–McLean Act of 1913. The act was very specifically directed at regulating the use of feathers in the fashion industry. It was shortly replaced by the more expansive Migratory Bird Treaty Act of 1918. Unlike the Lacey Act, which was directed at game bird species of economic and recreational value, the Weeks–McLean Act and subsequent Migratory Bird Treaty Act were directed at birds enjoyed for their aesthetic and ecological value. In a sense, the Migratory Bird Treaty Act was complementary to the Lacey Act; most birds in the United States that are not game species are migratory species and vice versa. Other than a few introduced or otherwise ubiquitous bird species such as European starlings, house sparrows, and rock doves, the Migratory Bird Treaty Act protects almost all nongame bird species in the United States. Even birds that are highly common and only weakly migratory over much of the United States, such as American robins (*Turdus migratorius*) and blue jays (*Cyanocitta cristata*), are protected under the act.

The word *Treaty* in the act is not an accident; the act was designed to serve as the statutory mechanism for enforcing what was in fact a treaty, executed between the United States and Canada (through Great Britain) in 1916 to protect birds that naturally move between the two countries. The treaty and the act were expanded in 1936 to include Mexico and later to encompass additional countries, such as Japan and Russia. The motivation for a treaty is rooted in the biology of many, indeed most, bird species in North America. Many North American birds spend much of the year in southern locales, usually the southern United States, Mexico, and Latin America, and fly north, usually to the central and northern United States, Canada, and Alaska, to breed. Flying north allows the birds to capitalize on the flush of vegetation and insect growth during the northern growing season during the time these resources are most needed (i.e., breeding) and then to retreat to the safety and availability of food resources in the southern locales once cold, barren winter weather returns to the breeding grounds. The need for international cooperation is therefore obvious; harmful actions in the locale of a portion of a species' life cycle can adversely affect the species everywhere. The need to include Mexico in the treaty was obvious by 1936; a need still exists to extend the treaty to the remainder of Central and South America.

The Migratory Bird Treaty Act, codified in its present form in 16 U.S.C. §§ 703–712, remains in place to the present day. The act provided the first protection to many of the marquee species later protected under the Endangered Species Act, including the bald eagle (*Haliaeetus leucocephalus*), peregrine falcon (*Falco peregrinus*), whooping crane (*Grus americana*), and wood stork (*Mycteria americana*). Recent removal (delisting) of the bald eagle and peregrine falcon from the Endangered Species Act following successful recovery of their populations has increased public awareness of the protections offered to these and other migratory birds by the Migratory Bird Treaty Act.

Central to the Migratory Bird Treaty Act is a prohibition on the ability to

> pursue, hunt, take, capture, kill, attempt to take, capture or kill, possess, offer for sale, sell, offer to purchase, purchase, deliver for shipment, ship, cause to be shipped, deliver for transportation, transport, cause to be transported, carry, or cause to be carried by any means whatever, receive for shipment, transportation or carriage, or export, at any time, or in any manner, any migratory bird, included in the terms of this Convention … for the protection of migratory birds … or any part, nest, or egg of any such bird.[7]

Note the use of the word *take*: The Endangered Species Act includes a similar prohibition on take of listed threatened or endangered species that is similar in some respects to the Migratory Bird Treaty Act's prohibition on take of migratory birds. Indeed, the Endangered Species Act includes the same prohibitions with respect to threatened or endangered species that the Migratory Bird Treaty Act extended much earlier to migratory birds.

Other key elements of the Endangered Species Act were foreshadowed in yet another earlier conservation statute, the Fish and Wildlife Coordination Act. As initially promulgated in 1934, the act authorized the secretaries of agriculture and commerce to assist and cooperate with federal and state agencies to protect, rear, stock, and increase the supply of game and fur-bearing animals, as well as to study the effects of domestic sewage, trade wastes, and other polluting substances on wildlife. As with the earlier Lacey Act, the focus was still on conservation of species of commercial and recreational value. The encouragement of cooperation among federal and state agencies was however new and has become a key hallmark of many more modern environmental acts, including but not limited to the Endangered Species Act. The Fish and Wildlife Coordination Act was modified in 1946 to require federal agencies to consult with the Fish and Wildlife Service and similar state agencies before authorizing, permitting, or licensing actions altering streams or other bodies of water. This consultation requirement, which remains in place to this day, foreshadowed a key element of the Endangered Species Act that requires consultation between federal agencies and the Fish and Wildlife Service before conducting, funding, or authorizing activities potentially affecting threatened or endangered species. Like the Lacey Act and Migratory Bird Treat Act, the Fish and Wildlife Coordination Act has experienced repeated amendments that altered or expanded its scope but remains in effect to the present day. In its present form, the act is codified in 16 U.S.C. §§ 661–667.

Even a closer foreshadowing of the Endangered Species Act pre-dating World War II was the Bald Eagle Protection Act of 1940, which was amended in 1978 also to encompass the golden eagle and presently codified in 16 U.S.C. § 668. The 1940 act prohibited taking, possession, and commerce in the bald eagle; the 1978 amendments extended these same prohibitions to the golden eagle. As the national symbol of the United States, the bald eagle has always stirred passions among the American people. The 1940 act extended protections reminiscent of those that would one day be imposed under the Endangered Species Act, but only to one species that the public held in exceptionally high esteem. Of course, as a migratory species, the bald eagle had received some protection under the earlier Migratory Bird Treaty Act, but the 1940 act enhanced and better publicized those restrictions. Despite the limitations placed on hunting and utilizing the bald eagle established in 1940, populations continued to decline precipitously, and once the Endangered Species Act was enacted in 1973, the bald eagle received protection as an endangered species (the then more abundant golden eagle was not listed).

For many years, the bald eagle was one of a few marquee species receiving the greatest attention under the Endangered Species Act, until visibly recovering populations led the U.S. Fish and Wildlife Service to upgrade the species' status to threatened in 1995 and eventually to delist it from the Endangered Species Act in 2007. The Service used the delisting to tout the success of the Endangered Species Act. The bald eagle was clearly the most publicly visible delisting under the act. The Service made it clearly known that many of the protections formerly afforded the bald eagle remained in effect through the Bald Eagle Protection Act (renamed since 1978 as the Bald and Golden Eagle Protection Act) and the Migratory Bird Species Act. In fact, the delisting did much to increase public awareness of these less publicly visible conservation laws.

The first direct predecessor to the Endangered Species Act was the Endangered Species Preservation Act of 1966. The 1966 act, amended in 1969 but ultimately repealed in 1973 on promulgation of the substantially broader Endangered Species Act, was a tepid effort to protect some of the most publicly visible species experiencing severe population declines. Listing was limited to vertebrates (i.e., animal species containing a backbone, such as mammals, birds, fish, and reptiles). The Endangered Species Act, when enacted in 1973, would encompass a broader scope, including not only vertebrates, but also plants, insects, shellfish, and other less-visible biological taxa. The Endangered Species Protection Act prohibited killing listed species but only in national wildlife refuges. The Endangered Species Act would later extend protection to listed species on all public and private lands, although it would extend stricter protections to federal lands and to actions sponsored by federal agencies. It would also regulate not only killing, but also adverse impacts to listed species, such as habitat destruction. The Endangered Species Act would also regulate how Americans handle declining species from foreign countries, while the scope of the earlier act was limited to domestic lands only.

The Endangered Species Preservation Act was only one product of the "environmental decade" of the 1960s. As explained by Eccleston and Doub,[8] the American public and Congress alike at that time were becoming increasingly concerned that the environment was deteriorating at an alarming rate. This was an era characterized by the publication of Rachel Carson's *Silent Spring*,[9] the Santa Barbara oil spill, and the Love Canal incident. Lake Erie was pronounced "dead," and smog alerts were issued in major cities across the nation. The Bureau of Reclamation was proposing to build a dam on the Colorado River that would flood the Grand Canyon. There were visibly polluted waterways, blighted urban landscapes, and unprecedented expansion of sprawling suburbs over the pastoral landscapes most visible to urban Americans. Events of this scope combined to stimulate an environmental activist movement.

Perhaps the crowning achievement of the 1960s environmental movement was enactment of the National Environmental Policy Act (NEPA) of 1969, presently codified as 42 U.S.C. §§ 4321–4347. Key among the provisions of NEPA was the following:

> All agencies of the Federal Government shall ... include in every recom-
> mendation or report on proposals for legislation and other major Federal
> actions significantly affecting the quality of the human environment, a
> detailed statement by the responsible official.[10]

That detailed statement became known as the environmental impact statement (EIS). Like so much of the earlier environmental legislation, such as
the Lacey Act, Fish and Wildlife Coordination Act, and Endangered Species
Preservation Act, NEPA was limited in scope to actions conducted by agencies of the federal government. As in previous decades, there was little
enthusiasm in Congress for imposing any but the most necessary prohibitions directly on the citizenry, and any such prohibitions could face serious
constitutional scrutiny. But forcing the federal government to consider the
environmental impacts of its actions before implementation was an unprecedented advance in environmental planning and conservation. Perhaps
NEPA's single greatest contribution has been that it requires federal agencies
to consider environmental issues in reaching decisions, just as these agencies consider other factors that fall within their domain. The bill's champion,
Senator Jackson of Washington State, declared that "no agency will be able to
maintain that it has no mandate or no requirement to consider environmental consequences of its actions."

Once implemented, NEPA became a key tool in the protection of species from extinction caused by actions of the federal government. One reason NEPA was such a trailblazing statute is its broad scope. Unlike most
prior environmental statutes, NEPA is not directed at any specific resource;
it applies to any issue "significantly affecting the human environment."
Agencies and courts rapidly recognized that the scope of NEPA includes consideration of impacts to fauna and flora. NEPA and the Endangered Species
Act, passed a mere four years later in 1973, have since played a closely interrelated role in the protection and management of faunal and floral resources
not only from actions directly sponsored by the federal government but also
from actions funded or permitted by the government. Thus, both NEPA and
the Endangered Species Act play a role in the planning of most major infrastructure development projects, including, among others, most newly proposed dams, freeways, airports, power generation facilities, and electric and
gas transmission facilities, whether directly sponsored by the federal government or proposed by privately owned companies receiving funding, permits, licenses, or other authorizations from the government. The Endangered
Species Act does, however, include provisions that do not overlap those of
NEPA, including provisions against the take of threatened or endangered
species that apply to nonfederal parties not regulated by NEPA.

Perhaps the most publicly visible products of the 1960s environmental
movement are the resource-specific acts commonly known as the Clean Air
Act of 1970 (42 U.S.C. § 7401) and the Clean Water Act of 1972 (33 U.S.C. §§
1251–1274). Unlike NEPA, and unlike the Endangered Species Act, these acts

directly regulate the actions of private corporations and citizens as well as the federal government. They empowered the Environmental Protection Agency (EPA), newly established in 1970, to establish thresholds for protecting the purity of air and water and to require permits for actions that cause emissions resulting in exceedances of those thresholds. Both acts, by virtue of their promotion of clean air and water, do of course indirectly help to protect the threatened and endangered species that inhabit those media.

But, the Clean Water Act is even more closely intertwined with the Endangered Species Act. Section 404 of the Clean Water Act, which regulates the discharge of "dredged or fill material" into "waters of the United States" has been interpreted by a series of court decisions to empower the federal government to require permits for most development projects that affect wetlands adjacent to rivers, streams, lakes, and other surface water features. Many threatened and endangered species listed or formerly listed under the Endangered Species Act are completely or partially dependent on wetland habitats. Notable examples include the bald eagle, whooping crane, wood stork, bog turtle, and swamp pink. The U.S. Army Corps of Engineers, which administers the Section 404 permitting program together with the older Rivers and Harbors Act permitting program, regularly coordinates with the U.S. Fish and Wildlife Service to ensure that the permits it issues comply with the objectives of the Endangered Species Act. Possible impacts to threatened or endangered species are a key consideration in many of the decisions made by the corps to issue or deny applications for Section 404 permits. Property owners and developers facing the delays and expenses of applying for Section 404 permits commonly mutter their disgust for the Clean Water Act and the Endangered Species Act in the same breath. A similarly intertwined implementation involves the Clean Water Act and the National Historic Preservation Act of 1966 (16 U.S.C. § 470), which requires reviews for impacts to historical and archaeological resources for actions sponsored, funded, permitted, or otherwise authorized by agencies of the federal government, including the U.S. Army Corps of Engineers.

1.5 The Endangered Species Act

When one considers the deeply rooted history of American conservation legislation protecting fauna and flora and the broad scope of environmental legislation passed in response to the 1960s environmental movement, the ultimate enactment in 1973 of the Endangered Species Act (7 U.S.C. § 136, 16 U.S.C. § 1531 et seq.) seems both logical and inevitable. Chapter 3 of this book examines the provisions and requirements of the 1973 act and its subsequent amendments in detail. The Endangered Species Act in a sense combines and strengthens elements of earlier conservation laws but targets them specifically to species formally designated as endangered or threatened (i.e., "listed") through a formal rule-making process that is open and responsive to public involvement and commenting. Similar to how the Lacey Act regulates commerce in game and other economically valuable species, the Endangered Species Act seeks to regulate commerce in listed species. Similar to how the Migratory Bird Treaty Act regulates take of migratory birds, the Endangered Species Act regulates take, by both federal and non-federal parties, of listed species. Like the Fish and Wildlife Coordination Act, the Endangered Species Act requires interagency consultation between federal agencies and the U.S. Fish and Wildlife Service.

Compared to most other conservation laws, the Endangered Species Act generally provides tighter, more rigorous protection and regulation of the listed species targeted for inclusion in the scope of the act. Inclusion of a species within the scope of the Endangered Species Act brings a vast arsenal of scrutiny to any actions that might affect that species. Simply the initial action of listing a new species often brings intense controversy to that species. After all, who other than a limited cadre of ornithologists and ecologists would know of the northern spotted owl (*Strix occidentalis caurina*), red-cockaded woodpecker (*Picoides borealis*), or whooping crane if those species were not included in the purview of the Endangered Species Act—even though killing individuals of those species would still be illegal under the Migratory Bird Treaty Act. The Migratory Bird Treaty Act, in some ways, is even stricter in its prohibitions than the Endangered Species Act. But, while nearly all Americans recognize the Endangered Species Act, few Americans not engaged in some form of professional environmental practice are even aware of the relatively obscure, and not highly controversial, Migratory Bird Treaty Act—or the Lacey Act or Fish and Wildlife Coordination Act. Even NEPA and the Clean Water Act are less widely known among the general populace than the Endangered Species Act.

The biological breadth of the Endangered Species Act is quite impressive. Perhaps among American environmental legislation, no act other than NEPA is as broad in its scope of applicability than the Endangered Species Act. Included among the species listed under the Endangered Species Act are birds, mammals, plants, fish, insects, reptiles, amphibians, and invertebrates—in short, if we are capable of losing something that lives, the scope of the Endangered Species Act can be extended to it. Of course, the United States does always have the political will to extend the protections of the Endangered Species Act to every species that might scientifically be in danger of imminent extinction; further chapters of this book explore some of the "listing controversies" in depth.

As of May 16, 2011, a total of 1374 species had been listed as threatened or endangered in the United States, including 582 animals (everything other than plants) and 792 plants. In addition to the strong protections offered by the act to species within the United States, it also as of May 16, 2011, extended protection to 595 designated threatened or endangered foreign species.[11]

Of course, the United States does not have the authority to impose many of the restrictions of the Endangered Species Act to species residing in foreign countries, but it can impose restrictions on commerce by Americans dealing in the species of foreign lands. Like the Lacey Act and Migratory Bird Treaty Acts discussed previously, the Endangered Species Act is not a purely domestic law.

1.6 Agencies Administering the Endangered Species Act

The agency with the most responsibility for administering and enforcing the Endangered Species Act is the U.S. Fish and Wildlife Service. However, recognizing that no single agency could have the requisite expertise to regulate such a broad spectrum of biota effectively, elements of the act involving species primarily inhabiting marine settings are administered by the National Marine Fisheries Service. The two agencies are frequently referred to as "the Services" in the context of the Endangered Species Act. Although it might seem logical to vest administration of the Endangered Species Act in a single lead agency, the joint approach takes advantage of the specialized expertise housed in each of the two biological service agencies. It also reflects the reality that the United States is a large and complex nation with a highly compartmentalized government, and the separate agencies of that government jealously guard their regulatory "turf." Of course, as discussed further in Chapter 4 of this book, the Services are not the only federal agencies involved in furthering the objectives of the Endangered Species Act. Indeed, all federal agencies have the responsibility of consulting with the Services before engaging in actions that could affect species covered under the Endangered Species Act.

1.7 International Protection of Endangered Species

While the United States was in the process of passing successively stronger species protection statutes that culminated in the 1973 passage of the Endangered Species Act, the decades of the 1960s and 1970s also saw growing international interest in the protection of rare species from extinction caused by exploitative trade practices. Visibly severe declines were taking place in the populations of game species valued in the ivory trade and by trophy hunters. A resolution to formulate an international agreement, termed the Convention on International Trade in Endangered Species of Wild Fauna and Flora (CITES), was adopted in 1963, agreed to by the signatory governments in 1973, and implemented in 1975.[12] Although CITES may be thought of as something of an international version of the Endangered Species Act, its scope is generally focused on the international trade and commerce in species, not broadly on conserving species. It may thus be better thought of as an international version of the Lacey Act, although even that comparison is substantially imperfect. Article II of CITES establishes "lists" of protected species somewhat similar to the lists of threatened and endangered species established under the Endangered Species Act. CITES itself is not a regulatory process; it instead requires signatory countries, of which the United States is one, to establish domestic regulatory processes that further the CITES goals of limiting exploitation of endangered species through trade. Indeed, the United States amended the Endangered Species Act to ensure that it includes the necessary provisions to comply with CITES.

CITES is discussed further in Chapter 3 of this book. However, this book is focused only on endangered species protection in the United States. Many countries have statutes protecting rare species; these statutes are conceptually similar to the Endangered Species Act, although their scope and rigor vary considerably. The United States' neighbor to the north, Canada, enacted its own Endangered Species Act in 1998 and amended it most recently in 2010.[13] The Canadian Endangered Species Act defines endangered and threatened species in a manner similar to the United States, but it collectively refers to those species, together with extinct and other special status species, as "species at risk." It prohibits many of the same actions as the United States, including actions that "kill, injure, possess, disturb, take or interfere with" endangered or threatened species. It requires the establishment of recovery plans for each listed species. Canada establishes as part of the statute a "Species at Risk Conservation Fund" to fund public activities to further the objectives of the act. In this manner, the Canadian act is reminiscent of the "Superfund" established directly under the U.S. Comprehensive Environmental Response, Compensation, and Liability Act to fund public cleanup of hazardous waste contamination sites. The United States neighbor to the south, Mexico, implements less-rigorous internal regulation of development activities affecting endangered species, but as a signatory to CITES and the Migratory Bird Treaty Act, closely regulates killing and exportation (and importation) of endangered species as designated under CITES.

Notes

1. Parks and Gardens Data Services. 2007. Deer Parks and Hunting—Parks and Hunting. Available at http://www.parksandgardens.ac.uk/274/explore-31/feature-articles-151/deer-parks-and-hunting-548.html?limit=1&limitstart=1. Accessed January 14, 2012.

2. King's Mark Resource Conservation and Development Project. 2011. Why the Name King's Mark? February 24. Available at http://ccrpa.org/km/King's%20 Mark%20Why%20The%20Name%20King's%20Mark.htm. Accessed January 14, 2012.

3. Eccleston, Charles H. and Doub, J. Peyton. 2012. *Preparing NEPA Environmental Assessments: A Users Guide to Best Professional Practices*, CRC Press, Boca Raton, FL.

4. Cornell Lab of Ornithology, 2003. *Handbook of Bird Biology*, Cornell University, Ithaca, NY.

5. 74 F.R. 59444–59472, Endangered and Threatened Wildlife and Plants; Removal of the Brown Pelican (*Pelecanus occidentalis*) from the Federal List of Endangered and Threatened Wildlife.

6. Nation Marks Lacey Act Centennial, 100 Years of Federal Wildlife Law Enforcement. U.S. Fish and Wildlife Service. May 30, 2000. Available at http://www.fws.gov/pacific/news/2000/2000-98.htm. Accessed May 6, 2011.

7. 16 U.S.C. 703.

8. Eccleston and Doub, *Preparing NEPA Environmental Assessments*.

9. Carson, Rachel. 1962. *Silent Spring*, Houghton Mifflin, Boston.

10. Sec.102(2)(C) of NEPA.

11. U.S. Fish and Wildlife Service. 2011. Summary of Listed Species Listed Populations and Recovery Plans as of Mon, 16 May. 01:48:16 UTC. Available at http://ecos.fws.gov/tess_public/pub/Boxscore.do. Accessed May 16, 2011.

12. Convention on International Trade in Endangered Species of Wild Fauna and Flora (CITES). Home page. Available at http://www.cites.org/. Accessed May 11, 2011.

13. Canadian Endangered Species Act of *1998, as amended, c. 11, s. 1.*

2

Some Basic Concepts

2.1 Introduction

Although the Endangered Species Act constitutes a national policy to protect biodiversity, promote species recovery, and prevent extinction, the act is rooted in science. Understanding the act requires some basic scientific knowledge. The Endangered Species Act is the federal environmental protection statute most rooted in biology, the scientific study of living things. More specifically, the Endangered Species Act is tightly rooted in the biological subdiscipline termed *ecology*, discussed in Section 2.2. Other biological disciplines such as botany, zoology, wildlife and fisheries conservation, entomology, and others play obvious roles. The scope of the Endangered Species Act is multidisciplinary in character. Its scientific foundation lies not only in ecology and many of the other subdisciplines of biology but also in physics, chemistry, mathematics, and planning and the environmental design arts. Table 2.1 presents examples of key scientific disciplines contributing to the technical knowledge underlying the Endangered Species Act and some of the corresponding species protected under the act.

2.2 Ecology

The general public commonly confuses the term *ecology* with environmental policy. For example, a group of environmental policy activists in Florida have formed a political party termed the Ecology Party of Florida, dedicated to "respect and reverence for, preservation and restoration of, the planet and its physical systems, for transparency and accountability at all levels of government."[1] Many people refer to themselves as *ecological* activists when in fact they are *environmental* activists. But in fact, ecology is a technical field of knowledge focused on the study of the relation of biological organisms to their physical environment. The American Ecological Society Web page presents the following definition for ecology:

Ecology is the scientific discipline that is concerned with the relationships between organisms and their past, present, and future environments. These relationships include physiological responses of individuals, structure and dynamics of populations, interactions among species, organization of biological communities, and processing of energy and matter in ecosystems.[2]

Notice that the definition describes ecology as a "scientific discipline." It is not a policy discipline. One need not be an environmental activist or have "proenvironment" or "green" political leanings to understand ecology or even be an ecologist; indeed, some ecologists are employed as subject matter experts by lobbyists and "progrowth" activist groups promoting eased environmental regulations. Less surprisingly, other ecologists work as subject matter activists for green groups and other proenvironmental activists. What these two disparate groups of subject matter experts, working toward sometimes-contradictory objectives, share is knowledge of a set of definitions, concepts, theories, and facts making up the subject of ecology.

More important, note use of the words *relationships, individuals, populations,* and *energy* in the definition of ecology. Most biological subdisciplines are concerned with specific groupings (taxa) of organisms, such as

TABLE 2.1

Scientific Disciplines Contributing to the Endangered Species Act

Discipline	Coverage	Examples of Listed Species
Mammology	Mammals	Black-footed ferret
		West Indian manatee
		Florida panther
Ornithology	Birds	Whooping crane
		Kirtland's warbler
		Wood stork
Herpetology	Reptiles and amphibians	Bog turtle
		Eastern indigo snake
		Kemps Ridley sea turtle
Botany	Plants	Swamp pink
		Eastern prairie fringed orchid
		Michaux's sumac
Ichthyology	Fish	Snail darter
		Robust redhorse
Entomology	Insects	American burying beetle
		Karner's blue butterfly
		Puritan tiger beetle
Invertebrate Biology	Invertebrates	Carolina heelsplitter
		Eastern rayed bean

animals (zoology), plants (botany), insects (entomology), fungi (mycology), and microorganisms (microbiology), or with specific elements of biological organisms, such as structure (anatomy), cells (cell biology), chemistry (biochemistry), and internal function (physiology). Ecology is broader: It involves the amalgamated consideration of multiple groups of organisms, elements of those organisms, and elements of the physical environment, such as water (hydrology), soil (soil science), and air (meteorology). Ecology is truly the big picture. Experts working to protect species and biodiversity often need (or must call on the expertise of experts in) the specialized scientific fields identified, but more than that, they really need to understand the big picture of how species interact with the physical environment. They need an understanding of ecology.

Endangered and threatened species, like all species, do not live in a vacuum. They live in the presence of other species. They feed on other species. Ecologists commonly speak of *food chains* in which plants and other photosynthetic organisms (autotrophic organisms, or producers) convert energy from the sun into biological tissue that can be used as food by other organisms (heterotrophic organisms, or consumers). Progressively larger consumer organisms then feed on smaller organisms through sequential stages referred to as *trophic* levels. The classic aquatic food chain image consists of a linear sequence of progressively larger fish, each preparing to swallow the fish just smaller than it, with the smallest fish feeding on algae or other submerged plant life. Terrestrial food chains might begin with plants, fed on by herbivorous (plant-eating) organisms such as meadow voles (*Microtus pennsylvanicus*), culminating with the mice being fed on by carnivorous (meat-eating) organisms such as red-tailed hawks (*Buteo jamaicensis*). These meat-eating organisms are commonly referred to as *predators*, and predatory birds are commonly referred to as *raptors*. Multiple interconnected terrestrial and aquatic food chains are usually present at any given natural setting, with the interconnected system referred to as a *food web*.

Species compete for food, resources, and space. Species live in a physical environment consisting of land, air, and water. They are subject to fires, floods, winds, precipitation, earthquakes, and mudslides. Scientists, politicians, and the American public may desire to preserve a species, designate it as endangered or threatened, and extend regulatory protections to it. However, they cannot remove that species from its biological and physical environment; they cannot parse individual species out. Species will persist only if they can continue to persist in their surroundings. In short, successful preservation of endangered or threatened species requires an understanding of ecology. Whether one works to advance the protection of threatened or endangered species or to ease the regulatory burden on industries whose work could affect threatened or endangered species, one needs knowledge of, or access to, experts with knowledge of, ecology.

2.3 Autecology and Synecology

The following paragraphs are not a comprehensive overview of ecology. They merely present some rudimentary concepts and terminology relevant to the protection and management of threatened and endangered species. Ecology is a complex science that cannot be readily summarized using a few terms, concepts, or mathematical formulae. Many ecologists divide ecology into two broad subdisciplines: one, termed *autecology* or population ecology, focuses on populations of organisms; the other, termed *synecology* or community ecology, focuses on the integrated function of populations of multiple species in spatial units termed *ecological communities*. Both are important to the Endangered Species Act. Endangered and threatened species live in habitats; protection and management of those habitats are essential to survival of the species harbored by those habitats. Populations of endangered and threatened species change over time; population changes that trend toward zero approach extinction. Persons managing threatened and endangered species are managing populations of organisms and the habitats in which they live.

2.3.1 Autecology

Autecology or population ecology addresses changes in the number of individuals of a species in a defined area, which can either be the whole world or a localized portion of the world. The term *population* is etymologically derived from the same Latin root as people and was initially used in studies of human numbers, but the term has been applied by analogy to plants, wildlife, and other species.

As a broad theory, populations of a species with access to unlimited resources will grow exponentially in a pattern that can be expressed mathematically as

$$N_t = N_0 e^{rt} \tag{2.1}$$

where N is the population size ($N_0 = N$ at start time, $N_t = N$ at time t), r is the intrinsic rate of growth, and t is time. Equation 2.1 is sometimes referred to as an exponential growth curve.

This is commonly explained using the example of bacteria placed on a fresh petri dish (glass or plastic container filled with growth medium). It can also be conceptualized as fish in a pond, deer in a forest, plants in a field, or birds on an island. Note that the rate of population increase is not constant (linear) but ever increases over time. The rate of population increase can be expressed mathematically using Equation 2.2, which is the equation for the first derivative of Equation 2.1:

$$dN/dt = rN \tag{2.2}$$

where N is population size, r is the intrinsic rate of growth, and t is time.

Population growth is not a linear equation with a constant slope (rate of change). Instead, the slope changes continuously as a function of its dependent variable N (population size), increasing as N increases. As the population increases, the rate at which the population grows also increases. The offspring in each successive generation breed in a similar pattern, so that successive generations contain ever more breeding individuals. Later generations therefore produce more offspring than preceding generations. Equation 2.2 is an equation for the slope as a function of where N stands.

The fact that populations of species increase at ever-increasing rates is evidenced by common observations of how weeds in a garden or algae in a pond can seem to explode from unnoticeable to overwhelming levels over unexpectedly short times. Based on Equations 2.1 and 2.2, an endangered species could recover, over time, if only endowed with optimum conditions. The species can be expected to grow in a pattern corresponding to these equations if protected not only from hunting or other human disturbance but also from natural environmental stress and resource scarcity. The fact that bald eagle populations expanded faster over the 1990s than in the 1970s[3] may also be reflective.

But, in the real world, the population of no species can grow exponentially indefinitely. Inevitably, some environmental resource, be it food, water, space, air, or whatever, will limit the exponential growth. The concept of environmental limitation can be expressed mathematically as

$$dN/dt = rN(1 - N/K) \qquad (2.3)$$

where N is the population size, r is the per capita rate of growth, t is time, and K is carrying capacity, a theoretical expression of the maximum population number that the subject area can support. Equation 2.3 is analogous to Equation 2.2, but it strives to account for limitations placed on the ultimate growth of a population posed by the inevitable limited supply of one or more necessary resources. When N is small relative to K, the population growth rate predicted by Equation 2.3 is very close to that predicted by Equation 2.2; as N approaches K (i.e., the population expands to approach carrying capacity), the population growth rate approaches zero. Equation 2.3 is sometimes termed the logistic growth equation. Population growth reflecting Equation 2.3 tends to grow in a sigmoid (S-shaped) pattern.

When confronted with a limiting resource, the rate of population growth increases over time and then decreases as the resource becomes more depleted. If the rate of population increase did not inflect, go from increasing to decreasing rates and ultimately level off, species would ultimately overwhelm their habitats. Petri dishes, lakes, forests, and other habitats simply can only accommodate so many individuals of any species or combination of species. The parameters r and K are properties inherent to each species and are fundamental in understanding population responses by each species to environmental limitations. Species designated as r-selected species tend to have fast reproductive rates and short life cycles. They tend to be near the bottom of food chains. They tend to be relatively nimble, recovering quickly once optimum conditions are restored following environmental stresses. K-selected species tend to have slower reproductive rates and longer life cycles. Young tend to require longer time to mature and reach reproductive age. They tend to be near the top of food chains. Populations of K-selected species tend to take longer to recover once limitations are removed or optimal conditions are restored.

Not surprisingly, most of the best-known threatened and endangered species are K-selected species. Bald eagles (*Haliaeetus leucocephalus*) did not immediately recover once protective regulations were established. Bald eagle females lay only one to three eggs per year, eggs require about 35 days to hatch, young require about 4 months to grow up, and individuals can live as long as 15 to 25 years.[4] This contrasts sharply with the life cycles of many bacteria and algae species, which can be measured in terms of hours rather than years. Recovery of the bald eagle was measured in decades, not even years. It was not on a scale comparable to algae reinfesting a pond or weeds reinfesting a field. Of course, a continuum exists between clearly r-selected species such as most algae and bacteria and clearly K-selected species such as most predators and raptors. Many species occupying intermediate stages on food chains display intermediate properties between r-selected and K-selected species.

One key limitation to population growth for a given species is competition for resources with other species. In the petri dish example of population growth, microbiologists generally inoculate (expose) petri dishes to a single species of bacterium or fungus. The inoculated species grows in a pure culture of only that species. Depletion of nutrients in the plate medium and space in the dish eventually limit the rate of population growth, but the limitations are experienced solely by the one species. In a field, forest, lake, or other natural setting, species usually exist in the company of other species and may compete with those other species for the same resources.

The populations of predator and raptor species tend to oscillate between rapid and slow rates of growth in response to rises and falls in the populations of the prey species that serve as their food sources. The oscillating predator and prey population levels can be expressed mathematically using equations termed the Lotka–Volterra equations. These equations, which are complicated, can be used to predict the responses of populations of predators to activities that alter their food supplies.

The competitive space occupied by a species is commonly termed by ecologists as its *niche*. A niche is more than just a spatial parameter. Niche encompasses not only the where and when of how a species is present, but also its overall behavior: how it feeds, reproduces, breathes, and raises its young and even how it dies (senesces). The more the niches of two species overlap, the more they compete. Two species competing for the same resource might have reached an equilibrium whereby each species steadily extracts a fraction of the resource, or the populations of each species might be responding to the presence of competition from the other, one population progressively extracting a greater proportion of a resource and the other extracting less. Ultimately, the population of the second species might decline, eventually to zero—extinction.

Populations of species can respond to resource limitations, competition, and other environmental stresses physiologically and reproductively. Progressively drier climate might induce deciduous tree species to shed leaves earlier in the season in an effort to utilize less water. If the drier climate persists over multiple generations of trees, it might favor the growth of individual seedlings more capable of conserving water. Water conservation properties can be expected to be distributed genetically over the population of seedlings based on probability theory (explained through the biological subdiscipline of genetics). However, the favored survival of water-conserving seedlings in the initial generations following the new stress can be expected ultimately to result in the increased presence of water conservation properties in successive generations. The drier conditions are said to select for water-conserving properties in successive generations.

The net response of the population, whether physiological, reproductive, or both, can result in a shift so that the niches of multiple species occupying the same setting overlap less. The genetic composition of populations can change in response to stresses (see Section 2.6), including stresses induced by resource scarcity resulting from competition. Populations of competing species can shift so that their niches overlap less and each experiences less stress brought on by resource scarcity. Ecologists recognize the tendency of multiple species sharing a habitat to gravitate to unique niches through a theory termed the *competitive exclusion principle*.

The ability to describe population changes mathematically using equations such as those presented enables ecologists to develop models predicting the possible responses of populations of threatened and endangered species to changes in environmental conditions. The use of mathematical models in ecology is conceptually very similar to the use of models in physics, economics, and other sciences. As in those other sciences, the numerical precision implied in the results of such models must not be interpreted as the ability to predict actual outcomes accurately. Casual reading of the business pages of any major newspaper will reveal numerous predictions of economic responses to various conditions: Will stock prices or stock averages rise, or will employment rise in response to a particular action proposed by the government? Separate columnists commonly disagree on the outcome of the same action, often in separate articles on the same newspaper page. And everyone has noticed that the actual economic results often—perhaps seemingly always—differ from what is predicted by a given columnist. Government scientists tasked with managing threatened and endangered species must make decisions regarding specific actions to protect those species, armed with ecological models but without knowing how those actions will actually affect the targeted species. Ecologists and economists approach their respective sciences facing many of the same conceptual uncertainties and using many of the same basic thought patterns.

2.3.2 Synecology

Synecology, or community ecology, focuses on the distribution of species across the physical environment. It focuses on habitats, distinct groupings of species occupying distinct spatial areas of land or water. Anyone who has visited a natural environment, whether as simple as a lawn or as complex as a forest or a wetland, knows that any environment comprises multiple species. If you think your lawn is completely bluegrass or fescue, look again—even the more intensively managed lawns are bound to include at least a few weeds. Inspection of most average lawns will likely reveal numerous weeds; plants such as common dandelion, large crabgrass, white clover, and dollarweed come to mind. Inspection of most forests will probably reveal multiple tree species as well as multiple sets of species occurring in visibly distinct horizontal layers termed strata. Strata can include a tree canopy, a shrub layer containing shrubs and tree saplings, and a low-growing ground-cover layer.

The groupings of plant species discussed are termed *plant communities*. A lawn is a simple plant community; forests, thickets, swamps, and marshes are more complex plant communities. In a nonecological sense, the term *community* might be applied to a neighborhood of butchers, bakers, candlestick makers, homemakers, students, children, retirees, and others. In an ecological sense, a plant community is a "neighborhood" of different trees, shrubs, wildflowers, grasses, ferns, mosses, and other plants. But, these natural neighborhoods contain more than just plants (flora). They usually also contain birds, mammals, insects, and other animals (fauna). Even more careful observation will usually indicate a less-visible but no less important complement of fungi, bacteria, and other microorganisms, the individuals of which often require a microscope to be seen.

Understanding habitats and possible effects of actions on habitats is essential to understanding effects on threatened and endangered species. Habitat is the home of a species; what affects the home affects the species. The Endangered Species Act recognizes the importance of habitat protection by offering direct protection to those habitats officially designated as "critical habitat" for one or more threatened or endangered species. Moreover, actions of federal agencies harming other habitats in a way that substantially harms listed species depending on those habitats can require an incidental take permit, even if the affected habitat is not designated as a critical habitat. However, nonfederal actions affecting habitats for threatened or endangered species, other than officially designated critical habitats, do not require a take permit under the act unless the effects clearly extend to the actual listed species.

Habitats have spatial and temporal (time) dimensions. Spatially, terrestrial habitats may appear to constitute two-dimensional surfaces that curve only with elevation but are actually three-dimensional volumes that extend upward and downward from the ground surface. Above the ground surface, terrestrial vegetation tends to be stratified, growing at multiple heights, where the leaves and stems of taller species may overlap and shadow shorter vegetation growing underneath. The soil beneath the ground surface contains not only plant roots but also soil-dwelling invertebrates such as earthworms and larvae of many insect species as well as fungi and other microorganisms that substantially influence surface soil structure and moisture and play key roles in decomposition of fallen leaves and other surface debris. Wildlife moves on the soil surface, but some wildlife, such as birds and bats, fly above the surface and often above the tree canopy; some other wildlife burrow into the soil to feed or shelter.

Habitats change in time by a process termed *succession*. Ecologists have traditionally recognized two types of succession: primary succession, whereby previously unvegetated settings develop vegetation, and secondary succession, whereby vegetation changes in composition. Some familiar examples of primary succession include the development of marsh grasses or mangrove trees on exposed tidal mudflats, development of beach grasses and wildflowers on newly accreting sand dunes, and development of mosses and grasses on exposed rock outcrops. Secondary succession is familiar to everyone who has witnessed the gradual transformation of a farm field to scrub and then forest or the reestablishment of forest in a clearcut left by timber harvesting. The distinction between primary succession and secondary succession is arbitrary; both involve the gradual transformation of natural settings through natural processes. Habitats are never static but are instead continually changing over time. Attempts to maintain lawns or meadows are attempts to arrest succession and perpetuate habitats in states that cannot persist without human intervention. The same is true for efforts to control weeds in fields or pastures or to cut hardwood trees out of planted pine stands. Keeping habitats in a state of rest requires continual application of force, much like pushing against a ball to keep it from rolling down a hill.

Succession is a continuous process. Any instantaneous state of a habitat during succession is termed a *sere*. Ecologists recognize that specific habitats in specific settings tend to change (succeed) in predictable patterns of one sere followed by another. The time interval that an individual sere persists can only be defined arbitrarily. Theoretically, succession constitutes continual progression along a continuous sequence of infinitesimally brief seres, but to be practical, ecologists recognize broadly distinct seres usually separated in terms of years. The theoretical conclusion of succession, the final expected sere, is termed the *climax*. Ecologists frequently speak of what the expected climax vegetation is for a given setting. Climax is thought of as a state of successional rest that persists naturally and indefinitely until the climax vegetation is disturbed by some external force, such as a storm, wild-fire, landslide, or disturbance by wildlife, wind, or human activity.

While climax is a useful concept, most ecologists recognize that it is only a theoretical concept that does not actually occur. While succession does indeed follow predictable sequences of seres in given settings, the process does not in fact stop at specific endpoints. The rate at which succession proceeds tends to be fastest immediately following disturbance and progressively slows over the subsequent years and decades. One may think of the successional process as tending to have negative acceleration. But, it never really stops. Succession is a response to changing environmental conditions, and environmental conditions are always changing. Scientists disagree regarding whether human activity is causing or accelerating global changes in climate, but few dispute that climate (like other environmental conditions) does change over time in response to natural factors. The few remaining stands of old-growth forest in the eastern United States that have never been cut over since European settlement (termed *virgin forests*) are several centuries old; their species composition may therefore reflect climatic conditions from past centuries rather than just those at the present. If those conditions substantially change over the next decades, the composition of these virgin forests can be expected to change even if they are never cut or cleared.

Many threatened and endangered species depend on specific successional seres. Well-known examples include the red-cockaded woodpecker (*Picoides borealis*) of southeastern pine forests and the Kirtland's warbler (*Dendroica kirtlandii*) of jack pine (*Pinus banksiana*) forests in Michigan. The red-cockaded woodpecker requires large, mature pines such as longleaf pine and slash pine to build nests.[5] Disturbed upland areas in the southeast typically progress over time from scrub to pine forests, but without new disturbance, the pine forests can be expected gradually to become hardwood forests. Prior to European settlement, many upland areas in the southeast experienced frequent wildfires, usually caused by lightning, that prevented succession to hardwoods. Not only did intense wildfires frequently "reset" succession, allowing new pine stands to grow, but also mature longleaf pines have thicker bark that is more resistant than that of hardwoods to brief wildfires. The wildfires essentially "weeded" the pine stands of hardwoods. The Kirtland's warbler requires a somewhat analogous situation involving stands of jack pine in north-central states, especially Michigan.[6] The jack pine stands of Michigan are highly analogous to southeastern pine forests; wildfires favor jack pine over hardwoods, and jack pine stands indefinitely protected from fire tend to convert to hardwood forests. Simply excluding development and timber harvest from red-cockaded woodpecker and Kirtland's warbler habitat will not ensure persistence of favorable habitat.

Ecologists have traditionally recognized three broad vegetative strata in forested habitats; in order from tallest to lowest, these are the tree canopy, the understory, and the ground cover. The understory typically consists of tree saplings (young trees) together with woody species that do not reach canopy height even when mature (commonly termed shrubs). The ground cover typically includes not only tree and shrub seedlings but also low-growing nonwoody plants such as wildflowers, ferns, and mosses. Nonwoody plants are commonly termed herbaceous and may include annual species, whose seeds germinate, grow to maturity, and release new seeds (complete their life cycle) over the course of a single year, and perennial species, which complete their life cycles over multiple years. In temperate landscapes such as the eastern United States, the top growth of many perennial species dies prior to winter, but the root systems persist over the winter and generate new top growth in the spring. However, some perennials such as Christmas fern (*Polystichum acrostichoides*) and most clubmosses (*Lycopodium* spp.) maintain live tops as well as roots throughout winter.

Reports prepared to document potential effects of actions on threatened or endangered species, such as biological assessments (see Chapter 5) or environmental impact statements (see Chapter 4) typically require maps and descriptions of habitats in the area of potential impact (commonly termed the action area). Boundaries between adjoining habitats may be discrete (abrupt) or gradual. Discrete boundaries can reflect abrupt changes in environmental conditions or past or present human activity, such as field or property lines. Most visitors to the seashore in the eastern United States are familiar with the visibly distinct belts of marsh grass vegetation in tidal marshes. These discrete habitat-type boundaries generally correspond to changes in tide elevations (water depth under specific tidal conditions). Visitors to forested tracts in former agricultural landscapes in the eastern United States commonly encounter abrupt changes in forest vegetation corresponding to historical farm field lines, abandoned roadbeds, or property lines.

Habitat boundaries are not always abrupt, however. Gradual changes in surface soil conditions, water table depth, or other environmental conditions can result in correspondingly gradual changes in vegetation composition defining the habitats. Even boundaries between wetland and nonwetland (upland) habitats can be gradual. The occurrence of such gradual boundaries is what makes wetland delineation so challenging. Land developers and their engineers who may be familiar with the abrupt visible boundaries between uplands and some wetlands, such as many tidal marshes, can become frustrated by delineated wetland boundaries that do not correspond to visible boundaries in the landscape. There may be a gradual boundary between two types of forested habitat. Both are recognizably distinct in their interiors, but the boundary lies within an interval of vegetation comprising elements of both.

Aerial photographs are a particularly useful tool for mapping habitat types on large tracts of undeveloped land, such as those favored by many threatened and endangered species. Discrete habitat boundaries generally reflect discrete changes in color, texture, and other photosignature elements in the photographs. But, drawing boundaries between adjoining habitats that intergrade gradually, usually both on the ground and with respect to photosignature on the aerial photographs, is as much art as science.

A key aspect for managing habitat is carrying capacity (the K factor in Equations 2.3 and 2.4). For a given habitat and species, the carrying capacity is the number of individuals of the species that the habitat can successfully support. If additional individuals enter the habitat, through movement or birth, an equivalent number will theoretically perish, as inadequate resources are available to support the increased number of individuals. Quantification of carrying capacity is difficult because most habitats support multiple species with overlapping resource requirements. The resource requirements of most species are not fully understood. Not every square foot of habitat, even if designated as the same habitat type, is identical. But, the concept of carrying capacity is critical to understanding how habitats can be managed to protect threatened and endangered species.

Several preserves in urbanizing areas in Florida contain specialized habitat required for the gopher tortoise (*Gopherus polyphemus*), a species listed by the state of Florida as threatened. Gopher tortoise populations in Florida are not federally listed under the Endangered Species Act,[7] but project proponents in Florida have taken actions for state-listed species that generally parallel those for federally listed species. Spontaneous, well-meaning attempts to relocate gopher tortoises to the preserve are tantamount to killing the relocated individuals. The preserve is at carrying capacity for the gopher tortoise; it lacks the resources to support any more. If an individual is relocated to the preserve, it will compete with existing individuals so that either it or an existing individual will perish.

Many environmental impact statements conclude that wildlife populations would not be significantly impacted by loss of a habitat because similar habitat adjoins or is located close to the project site. But such a conclusion must consider carrying capacity; displaced wildlife might move to nearby undisturbed habitat, but if the receiving habitats are at carrying capacity for those species, some combination of displaced or existing individuals would perish. Over the long term, the net effect is a population decrease.

2.4 Species and Taxonomy

Application of the principles of ecology, including autecology and synecology briefly discussed in Section 2.3, is necessary to provide a scientific foundation for implementing the Endangered Species Act. But, any understanding of the Endangered Species Act must also consider what exactly is a species. Everyone has an undefined conceptualization of the term *species*; every living thing is a part of some species. Dogs are dogs, cats are cats, dandelions are dandelions, copperheads are copperheads, and so forth. Within the broad array of dogs, we encounter features that when they occur together indicate some inherent "dogness": fur, four legs, one tail, and communication by barking. Even dogs recognize their own dogness; dogs clearly behave differently around other dogs than when they are around cats, horses, pigs, or other animals that may possess some of the features of dogness but lack the unique combination indicative of an actual dog. Even dogs that at a gross visual level look quite different, such as a golden retriever and a pug, still recognize each other as dogs and recognize cats as not being dogs. Specifically, dogs try to do something with other dogs that they absolutely refuse to try to do with cats or any other type of animal—they try to *breed* with other dogs. Even dogs as different as pugs and beagles try to breed; the offspring are marketed as "puggles." If you doubt the ability of widely divergently appearing dogs to breed, just consider the incredible array of "mixed breeds" in American dogdom today. They can all breed with each other. And none of them breed with anything else than another dog. Pugs, beagles, Great Danes, Chihuahuas, golden retrievers—they all belong to the dog species, *Canus familiaris.*[8]

The Endangered Species Act recognizes this ability to breed as the defining boundary to a species. It defines a species as any subspecies of fish or wildlife or plants and any distinct population segment of any species of vertebrate fish or wildlife that interbreeds when mature.[9] The act focuses not on protecting individuals but instead on protecting groupings of organisms capable of breeding with each other. Obviously, protecting those groupings involves protecting individuals making up the groupings. However, if there is no perceived threat to continued existence of a grouping, the act does not strive to promote more individuals of that grouping, regardless of the potential economic or ecological value of having more individuals (other regulations may however serve to promote more individuals of common species of economic or ecological value). Note also the use of the word *subspecies*; more on that in Section 2.5.

To better understand the concept of a species requires some basic understanding of the biological subdiscipline of taxonomy. Although humans have always attempted to describe the vast diversity of plants and animals that make up everyday life, the systematic approach to naming and cataloguing species used in modern science traces to the work of Carolus Linnaeus in the mid-1700s. Linnaeus is best known today not only for laying the foundations of modern taxonomy but also for promoting the use of the binomial Latin-based names for species. Commonly called "scientific names" or "Latin names," the binomial names assigned to each species consist of two words, a genus and a specific epithet. Consider the well-known endangered species with the common name "whooping crane" (pictured on the cover of this book). The scientific name is *Grus americana*. The name is italicized because it is an application of the Latin, not English, name, and Latin words are always italicized when used in English writing. *Grus* is the genus, which encompasses several similar species of tall wading birds, including the whooping crane, the more common but just as visually striking sandhill crane (*Grus canadensis*), and several similar species found only outside North America.

The specific epithet is *americana* (note that common practice is to capitalize the genus and not capitalize the specific epithet). The specific epithet, when combined with the genus, provides a unique identifying name for a single species. Note that the name of the species is the totality of the genus followed by the specific epithet, that is, *Grus americana*. The species is not *americana*, even though many improperly speak of the second component of the scientific names as the "species name" in an attempt to parallel use of the first component as the "genus name."

A not-uncommon misconception is that the scientific name of a species serves as an abbreviated description of the species. This is not true. The word *Grus* really conveys no information on the morphology (appearance), behavior, or distribution of the whooping crane. The specific epithet *americana* does suggest that the whooping crane occurs in North or South America, and that is true. But the whooping crane is not the only *Grus* species occurring in the Americas; the sandhill crane, *Grus canadensis*, is in fact far more common and widespread (and has always been so) in the Americas than the whooping crane. In fact, many specific epithets refer to the country or state where a species was first described. For example, the blackjack oak, *Quercus marilandica*, was first described based on specimens from Maryland, but in fact the species is more common in portions of many other southeastern states. To further lay the misconception to rest, consider that the term *whooping crane* actually conveys much more descriptive information about this imposing bird species with its distinctive whooping call than does the term *Grus americana*.

Another misconception is that the scientific name is less prone to change than the common name. Certainly, the scientific name is used identically in writing regardless of the language used. Unlike scientific names, common names must be translated into whatever language is being used. But, scientific names are hardly static. Consider the endangered eastern bog turtle (*Glypemys muhlenbergii*), a species inhabiting soft sediment wetlands in many eastern states. In some former literature, it is referred to as *Clemmys muhlenbergii*. In both the current and former literature, its common name is bog turtle. Possible reasons for changing scientific names are numerous but usually relate to new insights on classification of species, in recent years sometimes resulting from DNA sequencing. The former names are termed *synonyms*.

Linnaeus did more than provide a way to name species. He also developed a hierarchical way to classify species. We tend to apply hierarchical classifications, at least informally, to lots of things. Consider cars. Most car enthusiasts do more than recognize cars as Toyota Camrys or Dodge Chargers. If a Toyota Camry is a specific type (think species) of car, then Toyota is a somewhat broader category (think genus) based on the manufacturer of the Camry. But, we classify at yet broader (higher) levels. Toyota Camrys and Dodge Chargers are both sedans, sharing common characteristics that are not shared by groupings of vehicles classified as utility vehicles, minivans, or pickup trucks. And while sedans, utility vehicles, minivans, and pickup trucks may be driven by anyone with a state driver's license, special licenses are required to drive buses, heavy trucks, and other special vehicles—an even higher level of classification. Going higher, these disparate group of motor vehicles are subject to distinctive licensing and regulatory policies than are other vehicles such as bicycles, skateboards, and horse-drawn carriages (one might carry the analogy even further by considering the last as "endangered" and facing possible future extinction). And going even higher one might consider vehicles not designed for land transit such as boats, airplanes, and rockets. To sum, this analogy using vehicles has identified six useful levels of hierarchical classification: model, Camry; make, Toyota; form, sedan; licensing category, vehicles requiring state driver's license; powering mechanism, motor; mode, land transit.

If reading the preceding paragraph was fun, try the process using some other common everyday classifiable items (e.g., television programs, tools, furniture, electronic equipment, or office supplies). If you work in an office, examine your own filing system—be it ever so organized at all. In short, you will likely notice that at least subconsciously you classify things hierarchically. This is what taxonomists have done for species. Starting with Linnaeus and following sequential refinements to the present, modern taxonomy has identified the following hierarchical classification system:

Kingdom

Phylum

Class

Order

Family

Genus

Species

For Linnaeus, kingdom was easy—plant or animal. Animal, vegetable, or mineral—and minerals are nonliving and hence out of the process. Most naturalists at the time considered themselves zoologists or botanists. For his part, Linnaeus considered himself foremost a botanist, although his system was not limited to plants.

Kingdom started to become complicated once the microscope was invented and scientists began to observe the breathtaking diversity of microbial life and the complexity of microbial life, as well as the complexity of traditional animal and plant life at the cellular level. Originally, kingdom Animalia was reserved for the mobile[10] creatures commonly recognized as animals or insects. It included mammals (fur-bearing animals who suckle their young with milk), birds, reptiles (e.g., snakes and lizards), amphibians (e.g., frogs and toads), fish, and insects (and insect-like creatures such as spiders).

Everything else was relegated to kingdom Plantae. The plant kingdom includes the expected trees, shrubs, herbs, grasses, ferns and fernlike plants, and mosses and moss-like plants, nearly all of which are nonmotile (except during reproduction) and photosynthetic (which means that they derive their energy from sunlight rather than diet). But, it also traditionally included the vast array of plantlike and not so plantlike growths commonly referred to as algae and fungi, as well as the even vaster array of microscopic organisms. Although most nonspecialists think of algae as seaweeds, algae actually encompass a stunningly diverse array of startlingly different organisms that abound in water and soil. Some of the algae referred to as green algae are large, multicellular photosynthetic seaweeds that are both morphologically and biochemically similar to other plants. But, some other green algae species are microscopic cells or filaments of cells that even under a microscope really do not look like plants. These microscopic algal species can grow rapidly to produce thick masses, termed *blooms*, of organisms on the surface of water or other structures that impart a plantlike green color. Some green algae species actually comprise single cells or connected cells that "swim" using antennae-like protrusions called flagellae—one's imagination might be more inclined to view them as tiny swimming green animals than as tiny plants. With few exceptions, all algae species are photosynthetic, although the biochemistry of photosynthesis in many algae taxa differs from that of the familiar green land plants.

But, unlike actual plants, not all algae are green. Algae have traditionally been described using colorful terms such as the aforementioned green algae as well as brown algae, red algae, and others. Like green algae, the brown and red algae have macroscopic species of seaweeds as well as microscopic species that are invisible to human eyes except when they grow blooms of scum over the surface of waters or structures. The familiar green color of most land plants and "green algae" is caused by chlorophyll, a pigment that absorbs light to carry out photosynthesis. (Physicists recognize that chlorophyll is green because it preferentially reflects light with wavelengths corresponding to the green color while absorbing light with wavelengths just lower and higher than those for green.) Many photosynthetic algae use light-absorbing pigments other than chlorophyll that absorb different wavelengths of light and hence do not appear green.

Fungi are even weirder to the untrained eye: think otherworldly in appearance. Most familiar are the mushrooms and toadstools, those soft masses of whitish or gaudily colored tissue that are grown in cellars for gastronomic use (mushrooms) or that appear as discarded toys or litter amidst wet lawns or forest leaf duff (toadstools). Fear the unknown when observing these nasty little gremlins; they do not bite, but many wild mushrooms and toadstools have devastating poisons; the slightest nibble could result in a rapid, painful death. Less familiar are the microscopic fungi, many of which play key roles in decomposing dead plant and animal debris, causing plant or animal diseases, and leavening bread or fermenting juices (yeasts). Because large (macroscopic) fungi are immobile like plants, and because microscopic fungi look a lot like microscopic algae, all fungi have traditionally been considered plants.

However, fungi are not photosynthetic; they must absorb energy from their place of growth, such as decaying leaves, rotting fruit, or a chemical growth medium in a petri dish. The pigments in certain bright-colored mushrooms and toadstools are not photosynthetic but might play other ecological roles, such as deterring predators. Because fungi are not photosynthetic and play such fundamentally different roles in the environment, most scientists now recognize fungi as belonging to their own kingdom, kingdom Fungi.

Many mobile microorganisms were once viewed as little swimming plants or animals depending on whether they were photosynthetic. Many of these organisms swim through water using tail-like extensions termed flagellae. Flagellated microscopic photosynthetic organisms were considered algae and part of the plant kingdom; flagellated microscopic nonphotosynthetic organisms were termed *protozoa* and part of the animal kingdom. But both groups of organisms are now generally considered by biologists to comprise a separate kingdom, the protists (kingdom Protista).

Perhaps even more fundamentally different from plants and animals are bacteria and bacteria-like organisms. These organisms are comprised of cells (or one cell) considerably simpler in constitution than familiar plants and animals. Specifically, cells of these organisms lack a nucleus and other internal cellular features inside the external cell membrane. Biologists refer to organisms comprised of these cells as prokaryotic, while plant, animal, fungal, and protist cells are referred to as eukaryotic. Prokaryotic cells are also generally much smaller than eukaryotic cells, although the presence or absence of a nucleus, not size, is the discriminating characteristic. The most familiar prokaryote to most nonscientists, but by far not the most numerous, are a few disease-causing bacteria, such as the bacteria that cause bacterial pneumonia, strep throat, and salmonella food poisoning. But these represent only a handful of the numerous bacteria species that occur by the millions in nearly every environment. There are also photosynthetic prokaryotes, including the algal-like cyanobacteria, sometimes called blue-green algae, that can form scum-like growths on the surface of stagnant water. The prokaryotes are now commonly classified in their own kingdom, kingdom Monera, even though they, like the fungi, were once considered plants.

The five-kingdom taxonomy described is far from universally accepted among biologists. While few biologists still adhere to the original two-kingdom approach, considerable disagreement remains concerning which expanded kingdom approach is most appropriate. The issue is not especially relevant to the management of endangered species; even the most pessimistic ecologists do not expect possible extinction of entire kingdoms. The brief discussion of kingdoms in the preceding paragraphs is intended only to introduce readers to the incredible breadth of biodiversity that ranges far beyond the traditional plant and wildlife species most commonly considered in the context of the Endangered Species Act.

The groupings below kingdom but above species (i.e., phylum, class, order, family, and genus) are generally arbitrary divisions but seem obvious based on visible aspects of appearance (morphology), such as size, shape, color, texture, and so on. As for kingdoms, there is considerable disagreement among biologists regarding the classification of species into these intermediate taxonomic ranks. But, two individuals sharing any of these taxonomic ranks, even genus, cannot necessarily breed. Taxonomists strive to achieve a system in which individuals sharing progressively lower ranks are progressively more similar. According to this objective, two individuals classified in the same genus should be more similar than those sharing the same family, those of the same family should be more similar than those of the same order, and so forth. Historically, aside from the meaningful concept of interbreeding capability, "similarity" was a somewhat subjective consideration. In recent decades, as scientists achieved a greater understanding of the biochemistry of genetics (see Section 2.6), chemical differences in the makeup of genetic material in the cells of organisms has increasingly served as a more objective, measurable standard for taxonomic similarity and has prompted taxonomists to revise the taxonomic classification of many species.

The Endangered Species Act does not directly consider preservation of kingdoms or other taxonomic ranks higher than species. But, like the elements of the periodic table, are species the smallest divisible unit of taxonomy? Is species the most appropriate taxonomic rank to target regulatory protection? Or, should regulatory protection be tied to a more specific (notice the ironic word choice here) taxonomic rank such as subspecies or even individual populations of subspecies? Centuries of observation since Linnaeus have suggested that the ability of multiple organisms to breed may not be some inviolable metric for defining the lowest and most precise taxonomic rank. Groups of individuals capable of interbreeding may still bear notable differences in morphology (appearance) or biochemistry. Separate populations belonging to the same species but occupying habitats in different locations might occupy different ecological niches. Interestingly, physicists have also found that the protons, electrons, and neutrons making up the elements of the periodic table are themselves divisible into even smaller particles of matter, such as quarks. The Endangered Species Act might have too simplistic a name; understanding the actual objectives of the Endangered Species Act requires consideration of taxonomic divisions even narrower than species.

2.5 Subspecies

Taxonomy might suggest the occurrence of discrete groupings of organisms, but in fact the groupings we refer to as kingdoms, phyla, orders, and other taxa are actually arbitrarily assigned names to recognizable intervals on a continuum of biological diversity. Only very similar individuals are capable of interbreeding, but breeding individuals need not be completely identical.

Biologists have recognized separated populations of certain species, usually but not necessarily occurring in separate geographic locales, that while capable of interbreeding with other populations still display morphological, behavioral, or other physical differences. Various terms have been applied to these lower taxonomic rankings, but the most common at this time is "subspecies." The term *variety* remains, however, the most commonly used term for subspecies of plants.

If you still have a birding field guide published prior to the 1980s, you may find two separate species pictured: the myrtle warbler of the eastern United States and the Audubon's warbler of the western United States. However, these species have been observed to interbreed where their ranges overlap. Many ornithologists (bird biologists) therefore felt that the two species should be designated as a single species. The American Ornithological Union now recognizes only a single species, the yellow-rumped warbler (*Dendroica coronata*), with two subspecies: the myrtle subspecies (*Dendroica coronata coronata*) and the Audubon subspecies (*Dendroica coronata auduboni*). Many birders measuring their lifetime accomplishments in terms of a list of observed species (their "life list") suddenly found their life lists arbitrarily shortened. Of course, their actual observational experience was not shortened—they had still experienced the joy of observing two forms of this single species—but they had to accept that they had recognized two subspecies of one species rather than two individual species. Readers of this paragraph who are not birders probably find it to be a needless ramble, but trust me, many birders take issues like this very seriously.

Most taxonomic arguments and controversies are academic in character. Whether a narrow cadre of researchers and professors disagree about the class, order, family, or genus of a species has little or no economic or regulatory implications, even if the species is threatened or endangered. Consider white-tail deer (*Odocoileus virginianus*), common over most of the United States. These large antlered herbivores are familiar to most residents of rural and suburban United States. Visitors to some of the larger islands in the Florida Keys, a string of islands extending southwest from south Florida, may however notice some cute-looking dog-size deer, the key deer (*Odocoileus virginianus clavium*, note that *clavium* refers to the subspecies). Key deer are a localized population of white-tail deer occurring only in the Keys. They have greater tolerance of saltwater habitats than mainland white-tail deer and can make better use of mangrove forest and other habitats that are more prevalent on the Keys than over most of the mainland.[11] They are thought to be descended from white-tail deer from the mainland that wandered onto the islands. They can interbreed with white-tail deer from the mainland; they are part of the same species. But they are a visibly distinct subspecies. And unlike the robust numbers of white-tail deer on the mainland, the key deer subspecies is very rare—and listed as endangered. Their rareness reflects not only their localized range in a small area of specialized habitat on a few small islands, but also the fact that tourism and development have further reduced the amount of suitable habitat and subjected the already-small population to new stresses, such as vehicular collisions and disturbance from household dogs. In past decades, the petite deer were also hunted for their venison. But, unlike most mainland populations of white-tail deer, the always-small population of key deer occupying rugged habitat with only limited supplies of freshwater has not been nearly as resilient to hunting.

Although the population has increased since protections were introduced, the key deer subspecies remains in imminent danger of extinction. One might argue that extinction of a subspecies is not a true extinction. The death of the last key deer would not constitute extinction of the white-tail deer. Semantically, if extinction is defined as the loss of a species, this argument has merit. But, the loss of a subspecies such as the key deer is still a loss of biodiversity. If species, like other taxonomic divisions, is simply an identifier applied to an arbitrarily defined interval of biodiversity, then the concept of extinction can be equally well applied to a narrower interval of biodiversity such as subspecies. The key deer, like many subspecies, occur in localized portions of the overall species' range. The loss of the key deer subspecies would at least be the extirpation of the white-tail deer from the Florida Keys. For the habitats making up these small islands, the loss of the key deer sub-species still represents the vacating of an ecological niche, a severing of links in the food web of those habitats, and an overall diminution of the complex functional relationships constituting those habitats. Perhaps the loss can be better understood through the following consideration of genetics and natural selection.

2.6 Genetics and Natural Selection

Genetics, like ecology, is a broad-based biological discipline applicable across multiple taxa. This book does not strive to summarize even the fundamental principles of genetics, but a few genetic terms and concepts are essential to an understanding of threatened and endangered species. Gregor Mandel, a monk in the nineteenth century, recognized that certain features of garden pea plants were passed to progeny in quantitatively predictable patterns that can be explained using the mathematical processes of probability. Subsequent researchers discovered similar quantitative patterns for the distribution of features to progeny in other species and recognized that groups of features were linked by way of visible elements of cells, observable in microscopes, termed *chromosomes*. Crick and Watson discovered in the 1950s that the principal chemical molecule in chromosomes, deoxyribonucleic acid (DNA), consists of a ladder-like linear sequence of acids that can longitudinally split and regenerate matching sides of the ladder, thereby resulting in precise replication of information, reflected by strings of acids, from parent cells to new cells. Exchange of this same sort of information by combination of DNA from two parents explained how traits could be passed with predictable probability along lines of progeny.

The physiological function of every organism comprises thousands of biochemical chemical reactions. The traits characterizing that organism are the end product of complex chains of chemical reactions moderated by the presence of catalysts termed *enzymes*. Segments of acids in the strands of DNA in the chromosomes of each cell code for and enable the production of specific proteins required to support biochemical life processes. Particularly important are the proteins that combine to form enzymes that moderate biochemical reactions. The information contained in a segment of DNA coding for one or more enzymes responsible for the expression of some inheritable trait that can be passed to progeny is termed a *gene*. Specific physical characteristics of organisms such as eye or hair color in many animals and flower color or leaf shape in many plants are produced by genes passed along successive generations of offspring. Most people are familiar with plants such as flowering dogwoods (*Cornus florida*) for which some individuals produce white-colored flowers and others produce colored flowers; in many instances, this phenomenon reflects the presence or absence of genes calling for production of enzymes required for reactions generating pigments producing the color.

With some exceptions, each individual organism constitutes a unique combination of genes received in combination from two parents. We all know from everyday observation that not every dog, oak tree, robin, or bald eagle looks exactly alike. That is because each individual of those species has received some unique combination of genes—a very long string of acids along strands of DNA contained in the chromosomes of its cells—from its parents (usually two). The DNA of different individuals of any species represents a unique outcome from random intermixing of genetic information received from each parent.[12] There is some degree of genetic separation between differing individuals of the same species, even the same subspecies. Threatened and endangered species are no exception—not every whooping crane, red-cockaded woodpecker, or key deer looks identical.

The DNA possessed by separate individuals of the same species differs, but not so that the two individuals cannot interbreed. If the DNA is so different that two individuals cannot interbreed, then they are not of the same species, even if they are similar enough to be grouped in the same genus or higher-level taxonomic rank. Theoretically, the greater the difference, the higher (broader) is the level of common taxonomic rank shared by the individuals. Thus, for example, individuals of the same genus theoretically contain more similarity of genetic information in their DNA than do individuals of the same family but not the same genus. A continuous spectrum of genetic similarity exists, and the differences corresponding to each taxonomic rank are bounding points along a spectrum. Biologists have recognized genetic similarity sufficient to allow interbreeding as the threshold for classifying individuals in the same species. The degree of similarity for other taxonomic ranks is more arbitrary.

Until recently, taxonomic classification relied mostly on morphological characteristics (appearance) of organisms. With a greater understanding of DNA and the biochemical basis of genetics, biologists have increasingly used direct chemical analysis of DNA to support taxonomic classification.

Every individual of every species, whether or not threatened or endangered, or even rare, contributes to biodiversity. The loss of every individual even of common species is the loss of some unique combination of genetic makeup. The loss of the entire genetic makeup constituting a species is recognized as extinction of that species. The loss of the complete genetic makeup constituting a subspecies constitutes the extinction of the subspecies but not extinction of a species. However, the pool of genetic information contributing to the overall species is markedly reduced by the extinction of a subspecies. Even the loss of a group of individuals of a subspecies, without complete extinction of that subspecies, can be a substantial reduction in the genetic pool available to future generations of the subspecies (and the overall species). If subspecies, species, and other taxonomic ranks are but convenient thresholds on a continuous spectrum of genetic identity, then biodiversity losses can be more precisely thought of in terms of diminutions of a pool of genetic information that only by severity can result in the complete loss of a subspecies or species.

For the practical purposes of managing species and natural habitats, the pool of genetic material can be considered a nonrenewable resource. In this respect, the gene pool available to a given species may be considered to be much like oil or coal: Once it is burned, it is gone. The loss of genetic material is, for all practical purposes, irreversible. However, DNA replication, while remarkably efficient, is not perfect, and sometimes imperfections in the replication process result in the generation of new genetic information through a process termed *mutation*. If organisms containing mutated gene sequences can survive their environment, then they will introduce new genetic possibilities into the overall gene pool. But, mutation adds to the gene pool only very slowly. One could argue that exposure of dead plant or animal material to high pressures over long times could eventually add to our supply of coal or oil, but not in time to keep our economy from running out of these energy sources. Any additions to the gene pool through mutation are unlikely to be fast enough to save species stressed by development from extinction.

However, even when preserving biodiversity is viewed from the perspective of preserving genetic information rather than preserving discrete taxa, natural resource managers must confront the practical reality that protecting every individual of every species is neither practical nor achievable, and not even desirable. The loss of individual plants and animals[13] due to predation, competition, and other natural processes occurs continuously and cannot and should not be stopped. The question is how much of the genetic pool can be lost due to human interference in natural ecological processes. Natural resource managers also have to recognize the economic obligations of expanding human numbers for food, shelter, and a high standard of living. A nation can afford to focus on ambitions such as preserving biodiversity only when it enjoys an adequately high standard of living not to be preoccupied with basic survival, even though that survival may in fact be partially dependent on preserving biodiversity. Preserving biodiversity, whether through actions such as exclusion of agriculture or other development (e.g., mining or urbanization) from habitats, restricting hunting or other harvesting, reducing noise or habitat intrusions, or replanting or reconfiguring habitats, almost always involves economic trade-offs. These trade-offs can result in not only financial costs but also costs attributable to loss of freedom. Property owners and hunters deprived of freedoms to use land as they desire or to use firearms to hunt as they desire frequently complain that restrictions imposed to protect endangered species or other environmental resources constitute an infringement of rights expressed in the Fifth and Second Amendments to the U.S. Constitution.

When confronted with environmental stresses, individuals of a given species differ in their ability to withstand the stress, survive, and reproduce. Initially, the spectrum of resistance to a new stress can be expected to be distributed as a probability distribution. However, with progressive generations, the distribution can be expected increasingly to favor those traits most suited to resisting the stress. This response is termed *natural selection*. Natural selection occurs continuously in all species in all habitats, as populations continuously respond to new and changing environmental stresses. Its speed is theoretically proportional to the severity of emerging new stresses. The genetic response of wild populations of plants and animals to environmental stresses has been mathematically expressed through the Hardy-Weinberg principle, allowing for probabilistic modeling of responses of individual populations to hypothetical stresses. Application of the Hardy-Weinberg principle recognizes both the theoretical existence of a possible equilibrium in the absence of stress as well as the possibility of random slow changes in genetic composition, termed *genetic drift,* even in the absence of substantial stresses. Of course, an environment completely free of stress is largely theoretical and rarely encountered in practical natural resource management.

Most species perceived as "common" and not in noticeable danger of extinction, such as the mainland white-tail deer, possess a robust pool of genetic diversity that can be passed in a probabilistic manner from parents to offspring. New stresses drive natural selection, resulting in shifts in the genetic composition of robust species, but they do not usually result in extinction. Species perceived as "rare" and comprising fewer individuals, such as the whooping crane, may possess a more limited pool of genetic diversity that is less likely to produce through natural selection combinations of traits capable of successfully responding to new stresses. They may therefore be more susceptible to extinction following a new stress.

Of course, even seemingly robust populations of species (and subspecies) have limits in their ability to adapt to new stresses. The passenger pigeon and the Carolina parakeet had abundant populations over broad areas of North

America in the centuries preceding European settlement. After European settlement, the former was aggressively hunted for food, and the latter was aggressively hunted as a pest that fed on crops. Weak efforts to protect these species were too little and too late to stave off extinction. The extinction of the passenger pigeon in particular is something of a mystery and may be more attributable to large-scale habitat changes in the forested landscape of eastern North America than to hunting pressure. Aggressive hunting of the American bison for food and sport resulted in similarly severe population declines, but unlike the passenger pigeon, its population rebounded once the stress of overhunting was alleviated. Physicists speak of stresses causing strain on materials such that low levels of strain are elastic (reversible) but, once crossing a threshold, become inelastic (irreversible). The strains imposed by settlement and agriculture in North America on the passenger pigeon and Carolina parakeet were inelastic, but those placed on the American bison were apparently elastic.

2.7 Conclusion

The concepts of ecology, taxonomy, and genetics introduced at an elementary level in the preceding sections provide a framework for understanding species and extinction from a scientific perspective. The desire to prevent extinction of species that lies at the heart of the Endangered Species Act is more complicated than similar attempts to save representative examples of coins, stamps, furniture, or buildings. The latter are discrete, static, and easily definable units. Species, in contrast, are dynamic expressions of genetic material that are not discrete units and continuously change in response to changing environmental conditions. Preserving one, or even a small number, of individuals of a species—even if everyone agrees on what individuals constitute that species—is not enough to save that species from extinction. And even if that species is saved from extinction, loss of individuals from a species constitutes a loss to the genetic heritage making up that species. Natural resource managers are confronted with many seemingly simple but actually complex questions—what constitutes a species, how much of that species should be protected, and what are the trade-offs to our society and economy of enforcing protections to species and individuals of a species.

Notes

1. Ecology Party of Florida. 2009, January 22. Constitution of the Ecology Party. Available at http://www.ecologyparty.org/constitution.htm. Accessed September 21, 2011.
2. Ecological Society of America. 2011. About ESA. Available at http://www.esa. org/aboutesa/. Accessed September 20, 2011.
3. U.S. Fish and Wildlife Service. Chart and Table of Bald Eagle Breeding Pairs in Lower 48 States. Available at http://www.fws.gov/midwest/ eagle/population/chtofprs.html. Accessed November 30, 2011.
4. U.S. Fish and Wildlife Service, Midwest Region. Fact Sheet: Natural History, Ecology, and History of Recovery. Available at http://www.fws.gov/midwest/ eagle/recovery/biologue.html. Accessed December 2, 2011.
5. U.S. Fish and Wildlife Service. Southeast Region, Atlanta, Georgia, Recovery Plan for the Red-Cockaded Woodpecker, Second Revision, approved April 11, 1985, 296 pp.
6. U.S. Fish and Wildlife Service, Midwest Region. Kirtland's Warbler (*Dendroica kirtlandii*) Fact Sheet. Available at http://www.fws.gov/midwest/endangered/ birds/Kirtland/kiwafctsht.html. Accessed November 30, 2011.
7. Some gopher tortoise populations in other parts of the United States are federally listed. Remember that some species are federally listed in certain geographical areas of the United States but not in other areas.
8. Although *Canus familiaris* is the older and more recognizable scientific name for the domesticated dog among the general public, taxonomists generally recognize domesticated dogs and wild gray wolves, which are capable of interbreeding, as belonging to a single species, *Canus lupus*, of which the domesticated dog (all breeds and mixed breeds) belongs to a subspecies termed *Canus lupus familiaris*.
9. 16 U.S.C. 35.1532(C)(16).
10. Describing the animal kingdom as mobile is a simplification. For example, the plantlike corals that form forest-like stands in many shallow tropical waters are in fact animals, not plants. A more meaningful characteristic is the ability of plants to sustain themselves through photosynthesis, while animals must feed on other organisms. Biologists refer to plants as autotrophs (self-feeding) and animals as heterotrophs.
11. U.S. Fish and Wildlife Service, South Florida Field Office. 1999, May 19. Multi-Species Recovery Plan. Available at http://ecos.fws.gov/speciesProfile/ profile/speciesProfile.action?spcode=A003. Accessed November 2011.
12. The situation of identical littermates is not considered in this simplified discussion.
13. Or other biota, such as fungi and bacteria.

3

The Endangered Species Act: The Statute and the Regulations

3.1 Introduction

Before delving into a discussion of the specifics of the Endangered Species Act, it is useful to briefly review some basic elements of U.S. civics. Laws, whether environmental or otherwise, begin as bills in Congress. After passage by both chambers of Congress (i.e., the House of Representatives and the Senate), a bill is delivered to the president, who can either pass it into law or veto it. If vetoed, the bill may be again taken up by Congress in an attempt to override the veto. Whereas initial passage of a bill by the House of Representatives or Senate requires only a simple majority vote, overriding a veto requires a two-thirds majority vote in both chambers. The action of taking a bill successfully through this process and making it law constitutes promulgating a law.

What was to become the Endangered Species Act was introduced in the House of Representatives by its primary sponsor, Representative John Dingell of Michigan, and 24 cosponsors as Bill HR 37, the "Endangered Species Conservation Act."[1] Representative Dingell has represented the 15th Congressional District of Michigan, encompassing a portion of the Detroit metropolitan area, continuously from 1955 to the present. In addition to the Endangered Species Act, Representative Dingell was the primary sponsor of the National Environmental Policy Act[2] (NEPA) in the House and played a role in promoting several other environmental protection statutes that we take for granted today.[3] NEPA and many of these other environmental protection statutes are discussed in Chapter 4 of this book.

Once successfully promulgated, a bill becomes a public law and is published in the *United States Statutes at Large*. The Endangered Species Act, once passed by the 93rd Congress and signed into law by President Nixon in 1973, began its promulgated life as Public Law 93-205.[4] The Endangered Species Act, like many other long-standing environmental and nonenvironmental laws in the United States, has been frequently amended since its initial promulgation. Like the original act, each amendment was introduced as a bill in one of the two chambers of Congress, passed in both chambers, and signed into law by the president. Each amendment altered, added to, or deleted parts of the original statute. Many of the amendments are only minor or only administrative, but some have constituted significant changes to the act. For example, Public Law 98-364,[5] promulgated in 1984, changed how marine mammals are protected under the act. More controversial and more recent amendments have eased restrictions placed by private landowners and are discussed more thoroughly in further chapters of this book.

Once promulgated, each amendment to the Endangered Species Act became itself a separate law and was independently written into the *United States Statutes at Large* and assigned its own statutory citation number. Public Law 93-205 was published as 87 Stat. 884.[6] One could of course refer to the published statutes to review the contents of the act, but an accurate review would require reading not only the contents of the original law but also those of every amendment. For frequently amended laws such as the Endangered Species Act, reviewing all of the statutes is not only difficult but also fraught with the possibility of overlooking key amendments, thereby misinterpreting which elements of the act are currently in force. To alleviate this problem, Congress established the *United States Code*. When a new law is promulgated, it is written into an appropriate location in the *United States Code*, a process known as *codification*. The Endangered Species Act was codified as 16 U.S.C. 1531–1544 (Title 16 of the *United States Code*, Sections 1531 through 1544). Each time a new law is passed amending laws in the code, the changes are codified in the same location in the code. In this way, someone turning to 16 U.S.C. 1531 et seq. (in other words, starting at Section 1531, since amendments expanding the scope of a law can require adding new sections to the code) can find a full and up-to-date version of the act. However, technically, it is the statutes and not the code that carry the force of law.

Everyday speech sometimes does not distinguish statutes from regulations. But, there is a significant difference. Statutes are written by Congress; regulations are written by federal agencies. Rarely are lawmakers scientists or other technical experts. Once authorized to implement a statute, federal agencies, staffed by technical experts, must come up with specific rules for how they will administer the statute. These rules are termed *regulations*. Although the agencies write the regulations, they are answerable to the public. They must follow the procedures for rule making outlined in the Administrative Procedures Act. After writing proposed regulations, the agency experts publish the proposed regulatory text in the *Federal Register*, which is published daily. The *Federal Register* announcement invites the public to submit comments and establishes a deadline. After the deadlines, the agency reviews the comments, adjusts the text of the proposed regulation accordingly, and publishes the finalized regulations in the *Code of Federal Regulations*. They are then in force. Like the statutes, regulations are frequently amended as agencies respond not only to statutory amendments but also to court cases, Congressional pressure, public pressure, and the simple experience of lessons learned as regulations are put into practice in the "real world."

Multiple sets of regulations have been published by federal agencies assigned to implement and administer the Endangered Species Act. For example, regulations covering how the U.S. Fish and Wildlife Service (FWS) administers the Endangered Species Act are contained in 50 C.F.R. 17[7] (Title 17 of the *Code of Federal Regulations*, Section 402). Regulations covering requirements for interagency cooperation under the Endangered Species Act are published in 50 C.F.R. 402.[8] Although regulations are ultimately driven by their underlying statutes, it is more often the regulations than the actual statutes that outline limitations or requirements that directly affect the everyday lives of the public. Hence, public controversies are often directed more at the specific contents in regulations than at the broader statutory language. A common complaint is that while statutory language is prepared by elected politicians, regulatory language is usually prepared by agency staff ("unelected bureaucrats") not directly answerable to the electorate. However, the procedures for public comment on regulatory changes under the Administrative Procedures Act help to ensure answerability to public concerns. The ability of the regulated public to influence the process, through the election of politicians capable of changing the statutes and through court cases, enhances the checks and balances modulating the regulatory process.

The remainder of this chapter provides an overview of the statute and introduces some of the key definitions, terminology, and concepts essential to it. Much of what is introduced is explored in greater detail, using more specific examples, in further chapters.

3.2 Overview of the Statute

Like other statutes, environmental and otherwise, the Endangered Species Act is divided into "sections" containing compartmentalized language outlining distinct elements or accomplishing necessary administrative functions. The act, presented in its entirety in 16 U.S.C. 1531-1544, consists of 18 sections. Section 1 is merely a table of contents to the text of the act. Section 2 is a statement of findings, purposes, and policy for the act. Section 3 presents key definitions used in the act. Section 4 covers how species are selected for listing under the act. Section 5 authorizes agencies to acquire land to further the objectives of the act. Section 6 encourages cooperation between federal and state agencies.

Section 7 is one of the best-known sections of the act; it calls for interagency consultation between agencies proposing or facilitating projects and agencies with specialized knowledge of threatened and endangered species. Section 8 calls for international cooperation between the United States and foreign governments to pursue the objectives of the act. Sections 9, 10, and 11 are perhaps the most controversial sections of the act; Section 9 outlines the specific prohibitions contained in the act, Section 10 establishes a permitting program allowing exceptions by permit for those prohibitions, and Section 11 establishes enforcement procedures and penalties. Sections 9 through 11 are generally more controversial than Section 7 because while Section 7 applies only to federal agencies, Sections 9 through 11 apply to non-Federal interests and individuals as well. However, by influencing the actions of federal agencies affecting private property and interests, Section 7 can also be highly controversial.

Section 12 addresses specificities related to threatened and endangered plants. Section 13 outlines specific amendments to other environmental protection acts resulting from enactment of the Endangered Species Act. Section 14 repeals the Endangered Species Conservation Act of 1969,[9] the weaker predecessor to the 1973 act. Section 15 establishes appropriations for the initial years following enactment of the act; appropriations related to the act are now addressed annually as part of the normal federal budgeting process. Section 16 establishes the effective date of enactment. Section 17 establishes that no part of the act takes precedence over any more stringent requirements established under the related but earlier Marine Mammal Protection Act of 1972.[10] Section 18 requires federal agencies to report annually (for each fiscal year), for each listed species, on the costs of implementing the act.

3.3 Some Basic Definitions

All scientific specializations are built around some key definitions. Many of these definitions serve to delineate the scope of the specialization. For the study of endangered species, two definitions lie at the very heart: The first is the definition of *species*, discussed in Chapter 2; the second is the definition of *endangered*, discussed here. Not surprisingly, these two definitions lie at the nucleus of a statute termed the Endangered Species Act. Most statutes and regulations dealing with scientific issues, including most environmental statutes and regulations, lay out definitions for key items addressed in their scope. The Endangered Species Act is no exception. Section 3.3.1 presents the statutory definition for an "endangered" species, and the subsequent sections present other key definitions fundamental to the Endangered Species Act.

Some of the key terms were defined in Section 3 of the original text of the act. The following discussion, however, focuses on the codified definitions in 16 U.S.C. § 1532 and on the regulatory definitions.

3.3.1 Endangered

From a plain English perspective, one would expect an "endangered" species to be a species experiencing conditions that imperil its continued existence, or at least its continued vitality or importance in one or more contexts. Indeed, the statutory and regulatory definitions for an endangered species reflect this plain English expectation. The Endangered Species Act defines an endangered species as

> any species which is in danger of extinction throughout all or a signifi-cant portion of its range other than a species of the Class Insecta[11] deter-mined by the Secretary to constitute a pest whose protection under the provisions of this chapter would present an overwhelming and overrid-ing risk to man.[12]

Regulations issued by the Department of the Interior define an endangered species as one "that is in danger of extinction throughout all or a significant portion of its range."[13]

This definition is quite logical. It emphasizes that for a species to be listed as endangered, it must be in danger of extinction, at least over much of its range. From the initial implementation of the Endangered Species Act in 1973 until its downgrading to threatened in 1995, the bald eagle was per-ceived to be in danger of extinction over the contiguous 48 states but secure from extinction in Alaska. Hence the bald eagle was regulated as an endan-gered species in the lower 48 states but not in Alaska. Extirpation of the bald eagle from the Lower 48 was not perceived to be acceptable under the act, even if it were to persist in one remote state. American lotus (*Nelumbo lutea*), a very common plant in the marshes of the southeastern states, is found in a few limited areas in southeastern Michigan. The Michigan government is concerned that the species might readily be extirpated from Michigan with-out protection, enough so that the state protects it under state regulations. But, the prospect that this species might no longer persist in one state near the northern edge of its natural range is not a basis for listing this species as endangered under the Endangered Species Act.

The statutory definition is the same as the regulatory definition except for the quite logical exclusion in the statute for insects that are serious pests. Clearly, the framers of the Endangered Species Act did not want to see the act used to ensure the continued existence of clearly undesirable species, such as certain mosquito species or the boll weevil. Many ecologists would argue that the fact that an insect species is economically undesirable does not justify allowing its extinction. The fragile interconnection of food webs and predator/pest relationships, according to these ecologists, supports an approach of avoiding extinction of any species.

But, statutory and regulatory definitions are ultimately the work of poli-ticians, lawyers, and policy analysts, with input from technical specialists such as ecologists and biologists where appropriate. Ecologists and biolo-gists are not the final arbiters of the definitions, even if in an ideal world they would be. Of course, most of our most egregious pest insect species, such as the gypsy moth (*Lymantria dispar*), Mediterranean fruit fly (*Ceratitis capitata*), and emerald ash borer (*Agrilus planipennis*), are nonnative introduced species whose extirpation in (i.e., elimination from) North America would be ecologically as well as economically (and politically) desirable as long as the species remained extant (i.e., not extinct) in their homelands, where their interplay with local food webs prevents them from being pest species.

It is interesting that the act singles out insects but not undesirable microbes such as those that cause polio, measles, or mumps. Should anyone ever raise the act as an impediment to the elimination of such clearly detrimental species, courts will likely interpret that the act's protections do not apply to these species.

3.3.2 Threatened

One could easily conceive of an endangered species act whose scope simply encompassed a single category of species—those designated as endangered. Alas, few environmental statutes in the United States are so simple. Before relegating this complexity to the well-justified reputation of American politicians for indecisiveness, remember that environmental statutes address issues rooted in science. Science rarely allows for sharply delineated groupings. Using the definition of endangered presented previously, one should expect that there are varying shades of endangerment—a broad spectrum extending from clearly endangered species on which little doubt exists to borderline endangerment, which can be vigorously debated among experts with differing perspectives.

The framers of the Endangered Species Act recognized a need for some transitional category and introduced into the act's scope a second, less-severe, category of species faced with the prospect of extinction, but less imminently so. To extend at least some of the act's protection to these less-severely declining species at an earlier stage, before they face an imminent possibility of extinction, might head off the need for ever having to classify them as endangered. This would make for at least a partially proactive rather than reactive approach. Experienced practitioners under the Endangered Species Act have become conditioned to thinking of endangered and threatened species as a collective group; indeed, for unknown reasons the phrase "threatened and endangered species" rolls off their tongues more frequently than the more logically ordered "endangered and threatened species."

The act defines a threatened species as one "likely to become an endangered species within the foreseeable future throughout all or a significant portion of its range."[14] Like the definition of endangered, the definition for threatened comports with the expectations of plain English. The act clearly intended the designation of threatened to be one less severe than endangered. The definition of this lower-tier designation is closely tied in with the higher-tier designation. The broadness inherent in the definition of endangered therefore extends to the definition of threatened. But note the use of the word *foreseeable*; there has to be some palpable imminence to reaching the criteria for endangerment. The mere conceptualization of possibly becoming endangered is not a basis for assignment of the threatened status.

3.3.3 Proposed

Contrary to the perception of some, aloof panels of bureaucrats do not arbitrarily decide which species are endangered and which are threatened. Instead, experts employed by natural resource agencies (typically the U.S. FWS and National Marine Fisheries Service) review data and make recommendations regarding which species warrant those designations. Following those recommendations, the agencies follow the formal rule-making process of publishing notices in the *Federal Register* proposing recommendations that specific species receive the designations. The *Federal Register* notices can include only a recommendation for a single species or can include recommendations for multiple species; however, each species and its proposed designation must be identified by name. Other agencies and the public are then invited to comment on the proposal(s), and the agency decides whether to finalize a rule adopting the recommendation(s) only after reviewing and responding to the comments. The rule listing the species becomes effective only after being published in final form in a second *Federal Register* notice.

Once an agency publishes a recommendation for listing a species as endangered or threatened, that species is then designated as "proposed endangered" or "proposed threatened." The protections provided by the act for endangered or threatened species are not yet afforded to the proposed species; those protections are reserved only for species receiving the designations following review of comments and publication of a final rule in the *Federal Register*. One exception is that proposed endangered species that are already officially designated as threatened still receive the protections accorded under the act to threatened species. No species is afforded protection under the act unless it is the subject of a duly adopted final rule. Natural resource agencies commonly advise proponents of actions potentially regulated by the Endangered Species Act to consider proposed species only because the status of those species may be elevated to threatened or endangered imminently, before completion of the action. As soon as a species finally receives a threatened or endangered designation, it receives the corresponding protections prescribed under the act.

3.3.4 Candidate

The U.S. FWS and National Marine Fisheries Service assign experts to review pools of species regularly for possible listing as threatened or endangered. Inclusion of species in those pools is based on consideration of multiple lines of scientific evidence that populations of the species might be declining at rates that warrant protection under the act. Species in those pools are termed *candidate species*.

The term *candidate species* is more the product of regulations issued under the Endangered Species Act than it is of the act itself. The statute does not formally define the term *candidate species*. Nevertheless, it does recognize the concept of a pool of species that should be monitored for possible future inclusion as threatened or endangered. The act authorizes the secretary of the interior to provide financial assistance to states for "monitoring the status of candidate species within a State to prevent a significant risk to the well being of any such species."[15]

But, the term *candidate species* is really a logistical concept that agencies found necessary when developing regulatory procedures for implementing the act. Regulations issued by the Department of the Interior define a candidate species as "any species being considered by the Secretary for listing as an endangered or a threatened species, but not yet the subject of a proposed rule."[16] Many Endangered Species Act practitioners commonly think of candidate as something of a third or "C-level" tier below endangered and threatened, but in fact the act does not formally recognize the candidate status or offer any regulatory protection to species receiving that designation. The ability to focus monitoring efforts on a narrowed pool of species recognized as having greater potential for future extinction than the broader pool of common species is a useful tool to agency staff assigned to propose new species for listing, but the concept has little applicability to the regulated public. The act affords no protection to candidate species.

Agencies do commonly advise project proponents to consider possible effects on candidate species because of the potential that those species could become threatened or endangered in the future. In fact, the process for a candidate species to be proposed for listing and then ultimately receive listing under the act is typically very long—on the order of several years or decades. Unless one were to consider impacts from the life cycle of a very long-term project such as a power plant, evaluation of effects on candidate species is of little practical consequence. Still, the fact that sufficient scientific evidence exists that a species was declining enough to prompt experts to place it in a pool of candidates for possible listing suggests that consideration of impacts to that species might be essential to an evaluation of general ecological impacts, independent of the Endangered Species Act.

Literature prior to 1996 sometimes referred to three levels of candidate species. Species meeting the current definition of a candidate species were referred to as Candidate 1 species. Species for which the Services had some indication that future listing might be warranted but not enough to warrant Candidate 1 status were referred to as Candidate 2 species. Species that the Services had once designated as Candidate 1 or Candidate 2 but that no longer appeared to warrant future possible listing were referred to as Candidate 3 species. The Services limited use of the term *candidate* in 1996 only to what were then Candidate 1 species. Although the term *candidate*, regardless of level, never conferred regulatory protection or restrictions under the act, some mistakenly perceived that any candidate designation conferred protection.

Shortening the list not only limited the field of species that might be subject to that misunderstanding but also simplified the concept of candidate species. The change was made in the wake of the Contract with America, which had targeted the Endangered Species Act as a complicated process that infringed on personal liberties and private property rights.

3.3.5 Critical Habitat

Although the Endangered Species Act provides regulatory protection only for endangered or threatened species, it does offer protection to one other resource: habitat designated as critical for protection of endangered or threatened species. That habitat is referred to as critical habitat. In a broad context, *habitat* is an area within which species can live (see Chapter 2 of this book for a more detailed discussion of habitat). Thus, a forest that provides conditions suitable for occupation by the bald eagle (*Haliaeetus leucocephalus*) is said by ecologists to be habitat for the bald eagle. A stream within which the snail darter (*Percina tanasi*) can live is said to constitute habitat for the snail darter.

In the context of the Endangered Species Act, not all habitat potentially suitable for one or more endangered or threatened species constitutes critical habitat. In fact, the act specifically notes that "except in those circumstances determined by the Secretary, critical habitat shall not include the entire geographical area which can be occupied by [a] threatened or endangered species."[17] *Critical habitat* for a given endangered or threatened species is defined as

> specific areas within the geographical area occupied by the species, at the time it is listed in accordance with the provisions of section 1533 of this title, on which are found those physical or biological features essential to the conservation of the species, which may require special management considerations or protection; and specific areas outside the geographical area occupied by the species at the time it is listed in accordance with the provisions of section 1533 of this title, upon a determination by the Secretary that such areas are essential for the conservation of the species.[18]

Basically, habitat for a species must be essential to survival of the species to be designated as critical habitat. Furthermore, the habitat must be specifically identified in a duly proposed and passed rule, much as a species must be to become designated as endangered or threatened. If one spots a wood stork (*Mycteria americana*), an endangered species, in the marsh behind one's house in Florida, the marsh may be considered to be habitat for the wood stork. But, only if a federal agency has passed a rule formally designating that specific marsh as critical habitat for the wood stork does it constitute critical habitat for the wood stork. Unless formally designated as critical habitat, an action disturbing that marsh would not be regulated under the Endangered Species Act, unless of course it disturbs actual wood storks.

Note that critical habitat can only be designated in connection with one or more listed endangered or threatened species. Habitat cannot be designated as critical only on the basis that it is itself rare or of exceptionally high ecological value.

3.3.6 Delisting[19] and Downlisting: What the Act Seeks to Achieve

In the first decades after promulgation of the Endangered Species Act in 1973, the emphasis was on listing—making sure that the act's protections were extended to every species requiring them. The urgency of getting deserving species on the list remains, of course, but after nearly 40 years of implementing protections under the act it is not surprising that the public is expecting some tangible measure of success, evidenced most conspicuously by removal of species from the list. The process of removing species completely from the list is termed *delisting*, and like listing, delisting can be fraught with controversy. Delisting removes most of the act's protections from a species, although the act still requires the Services to monitor the continued recovery of delisted species to ensure that returning the species to listed status is not needed to achieve the act's objectives.

To many environmental activists, the prospect of delisting, while outwardly a declaration of victory, raises fears of a return to the practices that led to the population depletion that drove listing in the first place. To many regulated landowners and industries, the prospect of delisting offers the hope of future relaxed scrutiny and reduced land development costs. To some employed directly or indirectly by the Services, even if they agree with the scientific rationale for delisting, delisting means a loss of control over future actions affecting the species and the political specter of enduring the possible controversy involved in relisting the species should future conditions suggest a renewed need for the act's protections.

Delistings have so far been few and far between. The most publicly visible delisting was that of the bald eagle in 2007. The bald eagle had been listed as endangered from the outset of the act and then been "downlisted" from endangered to threatened in 1995. The publicly treasured species had become increasingly common and visible by the early 1990s, and Republican politicians had whipped up a frenzy of antagonism to government controls over private industry and landownership in the wake of the 1994 congressional elections and the Contract with America. Delisting the bald eagle was, of course, the objective of the more conservative politicians, at least of those who did not have outright repeal of the act in their sights. As things shook out, the neoconservative Republicans had to accept downlisting as a temporary compromise. Because threatened species receive most of the same protections under the act as do endangered species, downlisting a species from endangered to threatened is considerably less controversial than delisting a species.

Even as momentum toward repealing or liberalizing the act waned in the subsequent years, the increasing population levels of the bald eagle continued to become ever more visible in most areas of the country. Ultimately, the consensus among both the experts and the public was that if anything could recover under the act, then the bald eagle was it—and final delisting in 2007 was declared a victory by both supporters and opponents of the act. Perhaps no single species has been as intimately associated with the Endangered Species Act among the general public as the iconic and visually impressive bald eagle; declaring its successful recovery was perhaps the best conceivable way to generate positive press for the act.

Some other highly visible delisted species include the peregrine falcon (*Falco peregrinus*), a large and distinctive raptor common around cliffs and more recently adapted to use the exteriors of tall buildings, and the brown pelican (*Pelecanus occidentalis*), a large and distinctive piscivorous (fish-eating) bird now commonly observed along the southeastern Atlantic and Gulf coasts. The recovery of the bald eagle, peregrine falcon, and brown pelican may be attributable more to the disuse of the insecticide DDT since the early 1970s than to recovery efforts implemented directly under the Endangered Species Act. The American alligator (*Alligator mississippiensis*) was formerly listed as endangered under the Endangered Species Act but has become quite common throughout its former range in the southeastern United States. In fact, the American alligator is one of only a few crocodilian species in the world that is not presently in danger of extinction. The American alligator would have been delisted, but it is now designated as "threatened due to similarity of appearance" due to its close resemblance to the still-endangered American crocodile (*Crocodylus acutus*). Unlike most threatened species, however, American alligators may be legally hunted (with appropriate federal and state permits) and are even raised as livestock on "alligator farms." Alligator meat is common on restaurant menus in some parts of the southeast. The "threatened due to similarity of appearance" enables permitting requirements that help reduce the likelihood that pursuit of the common species results in inadvertent killing of the visually similar threatened or endangered species. But most other regulatory protections under the act are no longer extended to the American alligator.

3.3.7 Extinct: What the Act Seeks to Avoid

If delisting because of recovery is the ultimate objective of the act, there is a much darker version of delisting—that which occurs when a listed species becomes extinct. Interestingly, considering that avoidance of extinction is so central to the Endangered Species Act, there is no statutory or regulatory definition of extinction under the act. Of course, the intuitive definition obvious to nearly everyone, scientifically educated or not, is quite adequate: A species is extinct when it no longer exists. The IUCN (International Union for the Conservation of Nature) has developed a more scientifically elegant definition for its Red List biodiversity status program: A species (or other taxon) is extinct:

> when there is no reasonable doubt that the last individual has died. A taxon is presumed Extinct when exhaustive surveys in known and/ or expected habitat, at appropriate times (diurnal, seasonal, annual), throughout its historic range have failed to record an individual. Surveys should be over a time frame appropriate to the taxon's life cycle and life form.[20]

A brief three-page notice in the *Federal Register* on December 12, 1990,[21] had a chilling finality to it: It briefly presented scientific evidence that the dusky seaside sparrow (*Ammodramus maritimus nigrescens*) no longer exists and announced that it would be removed from listed status under the act as of January 11, 1991. Basically, the FWS decided that there was no reasonable doubt that the last dusky seaside sparrow had died. The Act failed, or at least the act lacked the ability to recover a species—actually a subspecies—whose population levels had dwindled so low following marsh draining actions in the early and mid-twentieth century that no practicable conservation measures were available to save it. None of us in our earthly lives will ever again witness a dusky seaside sparrow.

As noted in Chapter 2, one might argue that the loss of a subspecies such as the dusky seaside sparrow does not actually constitute extinction. Seaside sparrows as a species (*Ammodramus maritimus*) remain alive over much of the coastal eastern United States; in fact, the species is relatively common. But, the biodiversity of Florida's tidal marshes has been reduced by the loss of the dusky seaside sparrow; the portion of the gene pool that was formerly expressed by the dusky seaside sparrow exists no longer. While the overall function of the Florida tidal marshes formerly inhabited by the dusky seaside sparrow does not appear to have been noticeably impaired, scientists may never fully understand all elements of the unique ecological niche formerly occupied by the dusky seaside sparrow. On the other hand, could the money and other resources expended in the effort to save one relatively obscure subspecies have been better spent on other conservation efforts, such as preserving habitat used by other bird species, at a time when Florida land prices were cheaper than at present?

By the early 2000s, most biologists thought that it was only a matter of time before a similar *Federal Register* announcement would be issued for the ivory-billed woodpecker (*Campephilus principalis*), a visually stunning woodpecker that used to frequent old-growth wetland and riparian forests over much of the southeast. The ivory-billed woodpecker is not a subspecies of an otherwise common species; its loss would constitute extinction of an entire species. The last documented sightings were in a tract of old-growth swamp forest in the early 1940s prior to clearing of the trees to support the war effort. But, scattered anecdotal sightings continued in the subsequent decades.

These sightings were enough to prevent scientific authorities from considering the species extinct. As the decades passed, the general consensus continued to move toward acceptance of the fact that the species was extinct. Then, researchers with the Cornell Laboratory of Ornithology claimed in 2004 that they had spotted ivory-billed woodpeckers in a remote swamp in Arkansas. The claim garnered publicity far beyond the arcane scientific literature of ornithologists (bird experts) and wildlife scientists and even beyond the enthusiastic musings of birders (bird-watchers) and other passionate amateurs; it jumped to the national press. The putative rediscovery of a large and visually stunning species such as the ivory-billed woodpecker captured the fancy even of people with absolutely no interest in birds or wildlife. The drama and pathos surrounding the possible rediscovery of the ivory-billed woodpecker has been captured in an easily read, almost novel-like book.[22]

Despite the abundance of digital photography tools available in 2004, the researchers could not take definitive pictures of the birds they claimed to see. Furthermore, repeated visits to the same area by other researchers did not yield additional claims of spotting the species. So, the presence of the ivory-billed woodpecker in the twenty-first century remains more legend than fact. But, it has at least temporarily staved off a decision to declare the species extinct. At least hope remains for the ivory-billed woodpecker—some more than what remains for the dusky seaside sparrow. The ivory-billed woodpecker therefore remains an endangered, not extinct, species under the act. And the Services can use authority under the act to acquire and conserve habitat for the ivory-billed woodpecker, something that they cannot do for a species that is officially extinct.

The difficulties inherent in answering the seemingly simple question regarding whether a species is extinct can also be illustrated by the Eskimo curlew (*Numenius borealis*), a small migratory shorebird that once migrated in large numbers from wintering grounds in Latin America across the prairies of North America to breeding grounds in Alaska and northwest Canada, thence to eastern Canada before returning to the wintering grounds. As noted in the most recent five-year review by the FWS,[23] population levels crashed during the late nineteenth century due to a number of possible causes, including hunting, loss of prairie habitat along the migration route, and extinction of the Rocky Mountain grasshopper (*Melanoplus spretus*), an important food source. The last confirmed sighting was in 1963, but there have been 39 unconfirmed sightings since, most recently in 2006. Interestingly, the report cited an estimated probability that the Eskimo curlew remains extant as only 3×10^{-4} (where 0 indicates certain extinction and 1 indicates certain continued existence) based on physical evidence and even lower if the opinions of specialists are also factored in. However, some of the unconfirmed sightings have been by multiple "experienced birders," including a reported flock of 23 birds near Galveston, Texas, in 1981.[24]

The status of the ivory-billed woodpecker and Eskimo curlew may be effectively viewed from the perspective of Schrödinger's cat. In an attempt to explain the then-emerging science of quantum physics, Schrödinger developed an interesting thought experiment. The experiment consisted of thinking of placing a live cat in a closed container together with a flask containing poisonous gas timed to release at an unknown random time. Once the cat is placed with the flask in the container and the door sealed, the cat will remain alive until the unknown random time is reached, after which it will die. The observer cannot see inside the container and does not know when the random time is reached. Thus, from the observer's perspective at any time point after closing the door, the cat may be either alive or dead. There is no way for the observer to know which, unless of course the observer could break open the door.

The last remaining known habitat of the ivory-billed woodpecker was a tract of old growth swamp in Louisiana termed the Singer Tract. Groups of scientists and politicians petitioned the federal government to purchase the Singer Tract to preserve it as a national park or other preserve dedicated, at least in part, to saving the ivory-billed woodpecker. When the government failed to act to purchase the tract and its owner cut down the trees (clear-cut) in the swamp in the 1940s, it was like closing the door on the cat. Scientists since then have been unable to state definitively that the ivory-billed woodpecker is extant or extinct.

The inability to thoroughly observe every place within the dense and difficult-to-traverse old-growth swamp forests remaining in the southeastern United States is analogous to the inability of Schrödinger's observer to see into the box. The ivory-billed woodpecker may have become extinct in the 1940s, it may have gone extinct unbeknownst to humans at some point between the 1940s and the present, or unseen ivory-billed woodpeckers may still roam some remote areas of old-growth swamp. Extinction is rarely a prominently heralded event but instead usually occurs unobserved by humans at some undocumented time and place. The dusky seaside sparrow ranged over a small area of easily observable habitat; its extinction could therefore be announced with relative certainty. Any eventual future conclusion that the ivory-billed woodpecker has become extinct will be subject to considerably more uncertainty. Even if scientists continue to be unable to document living ivory-billed woodpeckers convincingly, anecdotal sightings will likely continue into the future.

The analogy can be similarly applied to the Eskimo curlew, although no one can identify any single action such as the loss of the Singer Tract that might have been the exact point of extinction. Thus, it is harder to speculate when the cat was placed in the chamber and the door closed. Nevertheless, despite the dogged attempts of enthusiastic birders to peer into the visually closed world of the Eskimo curlew to ascertain whether the species still lives, no one has succeeded. It may be extinct, or it may not be. Both the ivory-billed woodpecker and Eskimo curlew share another trait contributing to the mystique and legend regarding their possible continued existence: Both can be easily mistaken even by experienced birders for similar-appearing common species. The ivory-billed woodpecker is larger but visually resembles the common pileated woodpecker (*Dryocopus pileatus*), and the Eskimo curlew is smaller but visually resembles the common long-billed curlew (*Numenius americanus*) and whimbrel (*Numenius phaeopus*). The pictures in a field guide might suggest that the corresponding rare and common species can be easily distinguished by careful observation, but any birder can attest that visually distinguishing similar-appearing species seen at a distance or in low-light conditions can be extremely challenging.

The conundrum has considerable practical ramifications in the context of implementing the Endangered Species Act. Listing species as endangered or threatened empowers federal agencies to protect, establish, and manage habitat for the benefit of those species, review proposed actions of other federal agencies and suggest practices to benefit those species, and regulate certain private actions (e.g., hunting and collecting) to protect those species. Proactive responses such as protecting new habitat and improving other habitat are required only when the benefitting species is still extant. The recently completed recovery plan for the ivory-billed woodpecker calls for management actions to protect old-growth forest cover providing potentially suitable habitat for the woodpecker on federal lands, extending public ownership to additional potentially suitable habitat, and encouraging and helping private landowners to implement conservation measures favoring the woodpecker.[25] Even if the ivory-billed woodpecker is actually extinct (Schrödinger's cat is dead), these measures can be expected to benefit other extant species favoring old-growth swamp habitat in the southeastern United States, such as many neotropical migratory warblers. Populations of these other clearly extant species therefore benefit from the uncertainty over the actual condition of another possibly but not certainly extinct species, the ivory-billed woodpecker.

3.4 The Listing Process (Section 4 of the Act)

The Endangered Species Act does not authorize regulatory protections for any species that are not duly listed under the act as threatened or endangered. The listing process is established in Section 4 of the act. Opponents of the act commonly target the listing process because protections can only be enforced for a species (or area of critical habitat) after it is listed (or for critical habitat, designated). A highly visible recent controversy involves the proposed listing of the dunes sagebrush lizard (*Sceloporus arenicolus*), a small lizard occupying a small area of specialized habitat in southeastern New Mexico and west Texas.[26] Much of the affected area is owned or otherwise targeted for energy development, including oil and gas drilling and development of wind generation turbines. Energy development interests are fearful that listing the species will limit energy development opportunities in the area or impose significant additional costs, making energy development economically infeasible. Following the process prescribed under Section 4 of the act, the FWS issued a *Federal Register* notice proposing to list the species as endangered in December 2010[27] and, in response to significant opposition, reopened the comment period again in April 2011.[28] Further chapters of this book explore this and other listing controversies in greater detail.

3.4.1 Criteria for Listing

Because the protections offered by the Endangered Species Act are limited to those species formally "listed" under the act, one might expect the act to establish criteria for listing. Indeed it does, although not highly specific ones. With respect to criteria for listing a species as threatened or endangered, the act's statutory language states:

> The Secretary shall by regulation promulgated in accordance with subsection (b) of this section determine whether any species is an endangered species or a threatened species because of any of the following factors:
>
> (A) the present or threatened destruction, modification, or curtailment of its habitat or range;
> (B) overutilization for commercial, recreational, scientific, or educational purposes;
> (C) disease or predation;
> (D) the inadequacy of existing regulatory mechanisms; or
> (E) other natural or manmade factors affecting its continued existence.[29]

The regulatory language is virtually identical and provides little additional insight. The regulations issued by the FWS in 50 C.F.R. 424 state:

> A species shall be listed or reclassified if the Secretary determines, on the basis of the best scientific and commercial data available after conducting a review of the species' status, that the species is endangered or threatened because of any one or a combination of the following factors:
>
> (1) The present or threatened destruction, modification, or curtailment of its habitat or range;
> (2) Over utilization for commercial, recreational, scientific, or educational purposes;
> (3) Disease or predation;
> (4) The inadequacy of existing regulatory mechanisms; or
> (5) Other natural or manmade factors affecting its continued existence.[30]

Note that the criteria for listing, as established in the statutory and regulatory language, are not limited to human-made exploitation. Most listed (or formerly listed) species have, of course, met the criteria in large measure due to human activity—overhunting or overfishing (e.g., gray wolf); habitat loss (e.g., Florida panther); pesticide use (e.g., peregrine falcon); or lack of protection against human activity (e.g., piping plover). Some, such as the California condor and whooping crane, were never abundant and may have come close to extinction (or even gone extinct) even if never affected by human activity. In fact, human conservation efforts might possibly have even staged off a natural extinction for certain of these species. Of course, nearly all areas of the United States have been substantially influenced by human activity, and hence human activity has likely at least contributed to declines in probably all listed species.

Remember that merely meeting the scientific listing criteria does not qualify a species for protection under the act. Reputable and even renowned scientists may agree that a species deserves protection, perhaps even urgently to stave off extinction. But, to receive protection, a species meeting the criteria must actually be listed. Listing is typically a long process and is not completely free of political as well as scientific considerations. The fact that listings have tended to increase under Democratic administrations and decrease under Republican administrations attests to this political influence.

3.4.2 Process for Listing

The act empowers the secretary of the interior to develop a list of threatened and endangered species and to revise the list periodically. Its statutory language states:

> The Secretary of the Interior shall publish in the *Federal Register* a list of all species determined by him or the Secretary of Commerce to be endangered species and a list of all species determined by him or the Secretary of Commerce to be threatened species. Each list shall refer to the species contained therein by scientific and common name or names, if any, specify with respect to each such species over what portion of its range it is endangered or threatened, and specify any critical habitat within such range. The Secretary shall from time to time revise each list published under the authority of this subsection to reflect recent determinations, designations, and revisions made in accordance with subsections (a) and (b) of this section.[31]

Of course, the secretary of the interior does not individually write and issue the list. The Department of the Interior employs biologists and other professionals with the expertise needed to assess which species meet the act's criteria for listing. The Services may also from time to time hire other professionals on a contract basis to provide needed professional expertise. But, the list is still not exclusively the product of professionals. Publication in the *Federal Register* gives the public an opportunity to comment and provide input. Landowners and industries whose day-to-day activities and ability to make a profit may be hindered by extension of the act's prohibitions and requirements to a species proposed for listing have an opportunity to weigh in and express their concerns. While the department must respond to the comments it receives, the extent that the comments can sway a decision is the subject of considerable debate and controversy. As is generally true for most environmental regulations, lawsuits and the threat of lawsuits help to nudge the department from straying from the apparent intent of the act's statutory language.

Persons and parties concerned with the absence of a species from the act's scope need not, however, wait for the Department to initiate revisions to the list. They may at any time petition the Department to add a species to the list. The petition may be from a concerned member of the public but more often originates from activist groups or state or local governments that employ expert biologists. With respect to petitioning for listing, the statute states:

> To the maximum extent practicable, within 90 days after receiving the petition of an interested person under section 553 (e) of title 5, to add a species to, or to remove a species from, either of the lists published under subsection (c) of this section, the Secretary shall make a finding as to whether the petition presents substantial scientific or commercial information indicating that the petitioned action may be warranted. If such a petition is found to present such information, the Secretary shall promptly commence a review of the status of the species concerned. The Secretary shall promptly publish each finding made under this subparagraph in the Federal Register.[32]

On receipt of a petition, biologists and other technical specialists with the Services will review the technical merits of the petition and, based on their review, the secretary of the interior will decide whether the species addressed in the petition warrant proposal for listing. Either way, the secretary's decision is published in the *Federal Register* (remember that the Endangered Species Act tends to be implemented in a transparent manner, continually seeking input from the public). Ultimately, the process many culminate in the species becoming listed. But, while the process may be transparent and accessible to the public, it is rarely fast; typically, the petition review and listing process extends for years, if not decades. And unfortunately, while the Endangered Species Act process is generally slow, the process of extinction, or at least severe population decline, can be fast—just ask certain birdwatchers about the status of some of our rarer migratory birds that lack any status under the act.

3.4.3 Development of Recovery Criteria and a Recovery Plan

The goal of the Endangered Species Act is recovery, not listing, of species. As soon as a decision is made to list a species, the act encourages development of a plan of action to accomplish recovery, after which the species can be removed from the list. The act's statutory language provides a framework for developing recovery plans for listed species. It calls for the secretary of the interior to develop and implement plans, termed *recovery plans*, for the conservation and survival of listed species. The Act calls for recovery plans, to the maximum extent practicable, to:

- give priority to those endangered species or threatened species, without regard to taxonomic classification, that are most likely to benefit from such plans, particularly those species that are, or may be, in conflict with construction or other development projects or other forms of economic activity; and
- incorporate in each plan:
  - a description of such site-specific management actions as may be necessary to achieve the plan's goal for the conservation and survival of the species;
  - objective, measurable criteria which, when met, would result in a determination, in accordance with the provisions of this section, that the species be removed from the list; and
  - estimates of the time required and the cost to carry out those measures needed to achieve the plan's goal and to achieve intermediate steps toward that goal.[33]

Most recovery plans are elaborate documents. They typically begin with scholarly, in-depth physical descriptions of the subject species; discussions of the species' habits, habitats, behavior, and life cycles (sometimes referred to as the "natural history" of the species); and the factors thought to have contributed to the decline in populations. The plans then proceed to outline procedures for protecting the species and encouraging reattainment of stable population levels. Note that the goal of a recovery plan is not just protection of remaining individuals of the species; the objective of the act is actual recovery of the species. Recovery plans not only outline procedures for protecting the remaining population such as avoiding disturbance of habitat, but also outline procedures intended actually to grow the population. The latter might include federal acquisition of habitat, physical improvement of habitat, capturing individuals for captive breeding purposes, or establishment of specialized habitat.

Recovery plans include one other essential component: measurable criteria for ascertaining whether the species has been successfully recovered. Remember, the ultimate goal of the Endangered Species Act is not listing of species or even protection of species—it is recovery of species. Logically, achievement of recovery objectives requires a means of measuring whether those objectives have been met. Typical recovery objectives for most species are expressed in terms of attaining threshold population levels that are breeding at high enough rates to indicate that the populations are self-sustaining. Theoretically, once the measurable recovery goals are met, the species should be considered for possible delisting. Of course, in practice, the process for delisting is, like the process for listing, influenced by politically driven policy considerations as well as the act's intended scientific considerations.

3.5 Other Key Sections of the Act

From a compliance perspective, there are three key sections of the Endangered Species Act. Section 7 establishes a requirement for federal agencies to consult with the Services prior to taking actions that could adversely affect threatened or endangered species or critical habitat. The requirements of Section 7 apply only to federal agencies or to proponents of projects requiring permits from or receiving financial assistance from federal agencies. Section 9 establishes perhaps the most basic element of the act: It establishes prohibitions on adversely impacting threatened or endangered species or critical habitat. In the language of the act, Section 9 prohibits "taking" of listed species or habitat. Unlike Section 7, Section 9 applies to everyone, not just federal agencies or projects involving federal agencies. Section 10 establishes a permitting process for allowing exceptions to the prohibitions of Section 9. Permits under Section 10 are termed *take permits*; they allow for legally impacting listed resources if certain conditions are met. If Section 9 is the idealistic element of the Endangered Species Act, Section 10 is the act's nod to reality. Without Section 10, the remainder of the act could not practicably coexist in our modern society.

3.5.1 Section 7: The Government's Planning and Consultation Process

As outlined in the historical background to the Endangered Species Act presented in Chapter 1, predecessor laws to the act emphasized interagency cooperation between federal agencies proposing development actions and federal agencies authorized to manage and conserve natural resources. This requirement for interagency cooperation is carried into the Endangered Species Act by means of Section 7. The Endangered Species Act recognizes that the primary repository within the federal government of expertise pertaining to the management and conservation of threatened and endangered species lies with the Services. Although other agencies may possess significant natural resources expertise in their own staffs, Section 7 requires those agencies to tap into the larger body of specialized natural resources knowledge housed in the Services whenever proposing actions that could adversely affect resources protected under the act. The internal natural resources expertise of other agencies is by no means sidelined under the act; instead, the act encourages integrated teamwork between agency staff and the Services' experts.

Section 7 is a consultation, not a permitting, process. The process is not designed as a one-step effort whereby an action agency submits a single package of information to the Services and awaits a response allowing its proposed action to proceed. Instead, the Services encourage the development of an extended working relationship whereby action agencies make sequential contacts to the Services and incrementally incorporate feedback from the Services as the project design continues to mature. The initial contacts in the early part of the project design process are informal, whereby action agencies write informal letters to the Services describing project objectives and preliminary concepts, and the Services respond with relevant file data on the presence or absence of listed species and suggestions on how design concepts might be modified to avoid potential conflicts with listed species. If the Services recognize that the proposed action has little potential to adversely affect listed resources, they typically respond with a letter indicating that no further consultation effort is required.

Later contacts are typically more formal; action agencies submit more detailed project designs, results of site surveys for listed species, and evaluations of how actions could affect those species. For proposed federal construction projects, action agencies may have to submit detailed reports termed biological assessments. The length and detail of a biological assessment typically varies with the complexity of the subject action and its potential to affect listed resources adversely. But, most include information on the known or possible occurrence of listed species in areas that might be affected by the project, results of field studies for listed species if performed, descriptions of potentially affected listed species, detailed descriptions and interim construction drawings (blueprints) of the proposed action, and most importantly an evaluation of the potential affects of the action on listed species. Biological assessments typically conclude with the action agency's conclusions on how their action might affect each listed species in the project area. Those conclusions are typically either "no effect" or "may affect," and if the latter, are either "but would not adversely affect" or "and could adversely affect." Although that sentence presents the range of possible conclusions using wording encouraged by the Services,[34] some biological assessments prepared by some agencies might use somewhat different wording to express the same conceptual conclusions.

Of course, the Services do not automatically rubber stamp the conclusions offered by the action agencies. If the Services perceive the biological assessment to contain insufficient information, they may refuse to consult until the action agency adequately expands the biological assessment. The Services may accept the biological assessment and consult with the action agency, but they may find for a different conclusion than the one proposed by the action agency. Or, the Services may concur with the action agency's conclusions.

If the ultimate conclusions are "no effect" or "may affect but not likely adversely affect" for all listed species (and critical habitats) involved, the Services' concurrence letters conclude the Section 7 process. If the conclusion for one or more listed species or habitats is "may affect and may possibly adversely affect," then the Services must prepare a biological opinion containing recommendations for minimizing the adverse effects. Although action agencies are not strictly bound to comply with every element of a biological opinion, they generally do so unless there are compelling reasons not to. If an action agency elects not to follow a recommendation in a biological opinion and lacks a supportable justification, it risks being sued for failing to comply with the Endangered Species Act.

3.5.2 Section 9: You Cannot Knowingly Kill or Harm Listed Species

Section 9 is the core of the Endangered Species Act; it contains the enforceable prohibitions against actions that adversely affect listed threatened or endangered species or critical habitats. What Section 9 is best known for is that it prohibits take of listed species. Although the word *take* intuitively suggests shooting or otherwise killing one or more individuals of a listed species, whether for sport, harvest, or convenience, its use in the language of the Endangered Species Act is considerably broader. *Take* is defined by the act as an action "to harass, harm, pursue, shoot, wound, kill, trap, capture, or collect or attempt to engage in any such conduct."[35] Elements of take under this definition are not always intuitive. Simply walking up to a nest or rookery of a listed bird species or a den of a listed mammal species and making noise that might interfere with reproducing or rearing young could constitute take, even if no individuals are directly killed or injured, as it could be interpreted as harassment or harm. Similarly, simply attempting to use noise or some other deterrent to drive an individual of a listed species out of an area could be interpreted as take because it constitutes pursuit and might result in harm.

Like so much else in U.S. environmental law, what constitutes take is neither obvious nor clearly delineated. That might not be so serious a problem if the enforceable prohibitions of the act were limited to other federal agencies, as in Section 7. But Section 9 applies to the general public as well. Most people on the street with little technical knowledge of ecology or environmental policy would not be surprised that the Endangered Species Act prohibits shooting an endangered species. But, many are surprised that simply cutting trees down or operating heavy equipment near the nest of an endangered (or threatened) species could result in fines or imprisonment.

The regulations issued by the Department of the Interior seek to address some of the ambiguity surrounding the definition of take. They define *harm*, one of the less clearly defined forms of take, as including "significant habitat modification or degradation that results in death or injury to listed species by significantly impairing behavioral patterns such as breeding, feeding, or sheltering." They define *harass*, another less clearly defined form of take, as "actions that create the likelihood of injury to listed species to such an extent as to significantly disrupt normal behavioral patterns which include, but are not limited to, breeding, feeding, or sheltering." Both definitions include the word *significantly*; the act was never intended to prevent minor activities or activities having only a minor probability of disturbing listed species. Both definitions also focus on "breeding, feeding, or sheltering." The ability to breed—reproduce—is of course central to the continuation and recovery of a species. The ability to feed and find shelter is likewise essential to continued propagation of a species. Activities disturbing nesting sites, rookeries, or dens of listed species are clearly within the regulatory embrace of the act. Activities taking place in habitat that is only visited transiently by listed species are generally outside of the regulatory purview of the act. But, considerable grayness remains between such obviously extreme examples.

Less well known, except perhaps among international travelers, is that Section 9 also prohibits import, export, or commerce in listed species. As one might expect, it prohibits exporting listed species out of the United States. Clearly, removing individuals of a listed species from their habitat in the United States and relocating them to a foreign country is in conflict with the objectives of the act. Even if the individuals were to be well-cared for in a foreign land (not always a safe assumption), each individual removed is one less in its native habitat. The act also prohibits the export of products from listed species. For example, exporting oils obtained from listed whale species or the edible fins of listed shark species (valued in parts of Asia) is prohibited.

Section 9 also prohibits the import of listed species from foreign countries into the United States. Although the American public is most familiar with listed species occurring in the United States, many rare foreign species not naturally occurring in the United States are also listed under the act. One purpose of the customs process experienced by international travelers entering (or returning to) the United States is to inspect for possible violations of the Endangered Species Act. An acquaintance of mine inadvertently returned from a vacation to Russia with beluga caviar purchased legally in accordance with Russian law on the street in Moscow. On returning to the United States, customs agents confiscated the caviar and imposed a fine for violating the Endangered Species Act. The fish producing the caviar (fish eggs) was listed under the Endangered Species Act.

3.5.3 Section 10: But You Can Get a Permit to Do So

Clearly, the U.S. government frowns on killing or otherwise adversely affecting (i.e., "taking") listed species. So do much, if not most, of the American public; after all, the Congress that enacted the Endangered Species Act and the subsequent Congresses that have not acted to eliminate the Endangered Species Act have been and continue to be elected by and answerable to the American public. But, the framers of the act knew that their statute had to be implemented in the real world. In the real world, there are sometimes extraordinary circumstances under which the consequences of not killing or harming a listed species are worse than those of doing so. Although some environmentalists do not realize it, decisions to protect the environment, including those to protect threatened or endangered species, usually come at some cost to somebody, even if the net result is preferable to the citizenry at large. Rerouting a road around a location harboring endangered species may make the road longer, more expensive, and less convenient. Closing a forest to lumbering or a cove to fishing can cost jobs. Excluding home building from even a portion of a developer's property can result in fewer and more expensive homes. And in our democracy, these affected travelers, timber workers, fishers, home builders, and homebuyers share something very valuable with environmentalists: the right to vote.

The framers knew that if the Endangered Species Act were to work, and to work it would have to continue to be accepted by at least much of the citizenry, it would have to allow for at least some flexibility to allow the legally sanctioned killing or harm of listed species. That flexibility is offered in the form of take permits. Yes, it is possible to get permission to kill endangered species—as long as extenuating circumstances overwhelmingly justify it. No, getting a take permit is no mere formality, no mere zoning variance— think even more difficult than a wetlands permit. A take permit is something that authorizes someone to knowingly kill or injure one or more of the last remaining individuals of a species recognized as being in danger of extinction. They are not granted cavalierly.

The FWS issues three basic types of permits under the Endangered Species Act.[36] The first and best known is the incidental take permit. This permit authorizes take incidental to an otherwise-lawful activity. Shooting a listed species is not incidental take; building a subdivision in areas potentially inhabited by such species is. Considering the very broad definition of take regulated under the act, the availability of incidental take permits provides a very important necessary flexibility. Development and commerce are necessary to the American economy, as are private property rights. The Endangered Species Act was never intended to be a straitjacket, even in areas of the country harboring threatened and endangered species. Parties seeking an incidental take permit are required to submit a habitat conservation plan outlining measures to minimize and offset impacts authorized under the permit. The habitat conservation plan can be thought of as generally analogous to the better-known "wetland mitigation plans" that developers commonly prepare as offsets to wetland impacts authorized by permits under the Clean Water Act.

The second general type of permit is the enhancement of survival permit. The availability of these permits is a concession to private property owners, allowing the owners to implement conservation measures proactively in the present in exchange for guarantees allowing future development activities. The conservation measures can be spelled out as either a safe harbor agreement or a candidate conservation agreement with assurances. Although controversial, as these measures essentially exempt properties from future restrictions under the act, they offer additional flexibilities of great importance to property owners seeking certainty regarding future economic options for their real estate. The measures also establish proactive conservation efforts at an early stage, benefiting listed species immediately rather than only following implementation of development activities.

The third general type of permit is the recovery and interstate commerce permit. This is essentially a scientific collection permit. Unless you are a scientific researcher or engaged in some element of scientific research, you will never likely receive this type of permit. This type of permit does however offer necessary flexibility. Without it, most research into the biology and habits of listed species would be impossible, as that research typically involves handling of individuals or work in habitat falling within the definition of take.

3.6 Rare Species Designations Outside the Scope of the Endangered Species Act

Although the focus of this book is on species designated as endangered or threatened under the Endangered Species Act of the United States, other systems to classify species endangerment and likelihood of extinction are also in use in the United States and in other countries. State-level designations for rare species in state boundaries are discussed in greater detail in Chapter 7 of this book and therefore not discussed further here.

CITES: The Convention on International Trade in Endangered Species of Wild Fauna and Flora (CITES), introduced in Chapter 1 of this book, identifies species in need of conservation using three criteria:

- Species at most risk of extinction. Signatory countries to CITES prohibit international trade in specimens of these species except when the purpose of the import is not commercial.

- Species that are not necessarily at risk of extinction presently but might become so unless trade is closely controlled. Signatory countries to CITES require permits for export of these species but may allow import without a permit.

- Species included at the request of one or more signatory countries that already regulate trade in the species and that need the cooperation of other countries to prevent unsustainable or illegal exploitation.[37]

The relationship between the first two criteria somewhat but not exactly parallels that between endangered and threatened under the Endangered Species Act.

IUCN Red List: The International Union for Conservation of Nature (IUCN) established the Red List to accomplish the following goals:

- Identify and document those species most in need of conservation attention if global extinction rates are to be reduced; and
- Provide a global index of the state of change of biodiversity.[38]

Assignments to the Red List take place only following a peer review by at least two qualified specialists and are reviewed at least once every 10 years (every 5 years if possible). Since 2006, IUCN has recognized species in seven conservation status categories, within three principal groupings, whose definitions are presented in the following simplified form:[39]

Extinct conservation statuses

- Extinct (EX): Taxa for which there is no reasonable doubt that the last individual has died. IUCN presumes that a taxon is extinct when exhaustive surveys in known or expected habitat, at appropriate times (diurnal, seasonal, annual), throughout its historic range have failed to record an individual. Surveys should have been over a time frame appropriate to the taxon's life cycle and life form. Examples are the passenger pigeon and dusky seaside sparrow.
- Extinct in the wild (EW): Taxa known only to survive in cultivation, in captivity, or as a naturalized population (or populations) well outside the past range. IUCN presumes that a taxon is extinct in the wild when exhaustive surveys in known or expected habitat, at appropriate times (diurnal, seasonal, annual), throughout its historic range have failed to record an individual (as described for extinct). As for the extinct status, surveys should have been over a time frame appropriate to the taxon's life cycle and life form. Examples are the Hawaiian crow (*Corvus hawaiiensis*) and Wyoming toad (*Bufo baxteri*).

Threatened conservation statuses

- Critically endangered (CR): Taxa facing an extremely high risk of extinction in the wild. Examples are the leatherback sea turtle (*Dermochelys coriacea*) and California condor (*Gymnogyps californianus*).
- Endangered (EN): Taxa considered to be facing a very high risk of extinction in the wild. Examples are the Steller's sea lion (*Eumetopias jubatus*) and green sea turtle (*Chelonia mydas*).
- Vulnerable (VU): Taxa considered to be facing a high risk of extinction in the wild. Examples are the polar bear (*Ursus maritimus*) and great white shark (*Carcharodon carcharias*).

At lower risk conservation statuses

- Near threatened (NT): Taxa close to qualifying for, or likely to qualify for, a threatened category in the near future. Examples are the
American bison (*Bison bison*) and Virginia pine (*Pinus virginiana*).

- Least concern (LC): A taxon is classified as least concern when it
has been evaluated against the criteria and does not qualify for
CR, EN, VU, or NT. Widespread and abundant taxa are included
in this category.

Even though IUCN uses the terms *threatened* and *endangered*, it does not mean
the same as under the Endangered Species Act.

The word *threatened* in the NT conservation status can be especially
misleading, as many NT species are actually quite common. Designation
as NT rather than LC simply reflects some minimal evidence of possible
future risk—nothing even approaching the criteria for threatened under the
Endangered Species Act. Consider the two examples of NT species noted.
American bison experienced breathtaking population drops and without
conservation measures could have gone the way of the passenger pigeon,
which was likewise a once very common species subjected to reckless hunting and harassment for both meat and sport. However, American bison are
now raised domestically and frequently serve as a lean alternative to beef in
restaurants. Virginia pine is common in early successional forests on former
agricultural lands throughout much of the eastern United States, especially
in the Piedmont and Appalachian areas. In many places, it is a dominant
canopy tree. It is also widely cultivated for use as Christmas trees.

United States WatchList (Birds Only): The United States WatchList is
a joint project between the American Bird Conservancy and the National
Audubon Society to identify those bird species in greatest need of immediate conservation attention. The watch list identifies two categories of species: red species, which are of greatest concern, and yellow species, which
are still of conservation concern but not as much so as the red species.
The designations are based on a review of data from the American Bird
Conservancy's Partners in Flight Program, the National Audubon Society's
Christmas Bird Counts, and the FWS's Breeding Bird Survey.[40] Because new
data are collected annually under these programs, by numerous professionals and amateur enthusiasts, in large numbers of locations throughout the
United States, the watch list is responsive to rapid changes in population
levels for bird species.

As of 2007, the watch list contained 210 species, including 93 red species
and 117 yellow species. The 93 red species included 55 species of the con-
tinental United States and 38 Hawaiian species. The yellow species are not
separated into subgroups based on geography. Many, but not all, of the red
species are also endangered and threatened. Some examples of red species that are not listed as endangered or threatened include the black rail

(*Laterallus jamaicensis*), reddish egret (*Egretta rufescens*), golden-winged warbler (*Vermivora chrysoptera*), and tricolored blackbird (*Agelaius tricolor*). Most of the yellow species are not endangered or threatened, except for a few threatened and even a couple of endangered species.[41] A few bird species listed under the Endangered Species Act, such as the wood stork, are neither red nor yellow.

Notes

1. Bill Summary and Status, 93rd Congress (1973–1974), H.R. 37. Available at http://thomas.loc.gov/cgi-bin/bdquery/z?d093:HR00037. Accessed May 18, 2011.
2. National Environmental Policy Act of 1969, as amended (42 U.S.C. 4321 et seq.).
3. Official Web site for Representative John Dingell, http://dingell.house.gov.
4. Public Law 93-205, Endangered Species Act of 1973, approved December 28, 1973.
5. Public Law 98-364, An Act to Authorize Appropriations to Carry Out the Marine Mammal Protection Act of 1972, for Fiscal Years 1985 through 1988, and for Other Purposes, approved July 17, 1984.
6. 87 Stat. 884, Endangered Species Act of 1973, approved December 28, 1973.
7. 50 C.F.R. 17, endangered and threatened wildlife and plants.
8. 50 C.F.R. 402, Interagency Cooperation—Endangered Species Act of 1973.
9. Endangered Species Conservation Act of 1969, as amended, 16 U.S.C. 668.
10. Marine Mammal Protection Act of 1972, as amended, 16 U.S.C. §§ 1361–1421.
11. The class Insecta comprises the various genera and species of six-legged fauna commonly referred to as "insects" but does not include the similar eight-legged fauna commonly referred to as "spiders" (class Arachnida). The insects, spiders, crustaceans, and other similar taxa are all part of the phylum Arthropoda (arthropods).
12. 16 U.S.C. § 1532(6).
13. 50 C.F.R. 424.02(e).
14. 16 U.S.C. 1532(20).
15. 16 U.S.C. 1535(d)(1)(F).
16. 50 C.F.R. 424.02 (b).
17. 16 U.S.C. 1532(5)(C).
18. 16 U.S.C. 1532(5)(A).
19. The discussion of delisting in the text presented here focuses on the desirable situation in which a species is delisted because its population has recovered. It does not focus on the unfortunate situation in which a species is delisted because it has become extinct.
20. About the IUCN Red List. Available at http://www.iucn.org/about/work/programmes/species/red_list/about_the_red_list/. Accessed August 8, 2011.
21. 55 F.R. 239, 51112–51114.
22. Gallagher, T. 2005. *The Grail Bird*, Houghton Mifflin, Boston, 272 pp.

23. U.S. Fish and Wildlife Service. 2011. Eskimo Curlew (*Numenius borealis*), 5-Year Review: Summary and Evaluation. Fairbanks Fish and Wildlife Field Office, Fairbanks, Alaska, August 31, 14 pp.

24. MacAlister, W. H., and M. McAlister. 2006. *Guidebook to the Arkansas National Wildlife Refuge*, Mince Country Press, Victoria, TX, 336 pp.

25. U.S. Fish and Wildlife Service. 2010. *Recovery Plan for the Ivory-billed Woodpecker* (Campephilus principalis), Southeast Region, Atlanta, GA, April 16, 156 pp.

26. U.S. Fish and Wildlife Service. 2011. *Species Profile for Dunes Sagebrush Lizard*, U.S. Fish and Wildlife Service, Washington, DC, August 6.

27. 75 F.R. 77801–77817, Endangered and Threatened Wildlife and Plants; Endangered Status for Dunes Sagebrush Lizard.

28. 76 F.R. 19304–19305, Endangered and Threatened Wildlife and Plants; Endangered Status for Dunes Sagebrush Lizard (reopening of comment period).

29. 16 U.S.C. 1533(a)(1).

30. 50 C.F.R. 424.11(c).

31. 16 U.S.C. 1533(c)(1).

32. 16 U.S.C. 1533 (b)(3)(A).

33. 16 U.S.C. 1531(f)(1).

34. U.S. Fish and Wildlife Service and National Marine Fisheries Service. 1998. *Endangered Species Consultation Handbook, Procedures for Conducting Consultation and Conference Activities under Section 7 of the Endangered Species Act*, March, Final. Washington, DC: Author.

35. 16 U.S.C. § 1532 (19).

36. U.S. Fish and Wildlife Service Endangered Species Program: Permits. Available at http://www.fws.gov/endangered/permits/index.html. Accessed August 7, 2011.

37. Convention on International Trade in Endangered Species of Wild Fauna and Flora (CITES). Home page. Available at http://www.cites.org/. Accessed January 6, 2011.

38. International Union for Conservation of Nature (IUCN). 2011. The IUCN Red List of Threatened Species, Red List Overview. Available at http://www.iucnredlist.org/about/red-list-overview#redlist_authorities. Accessed January 6, 2012.

39. International Union for Conservation of Nature (IUCN). 2011. About the IUCN Red List. Available at http://www.iucn.org/about/work/programmes/species/red_list/about_the_red_list/. Page last updated May 24, 2011. Accessed January 6, 2012. For clarity, the definitions presented in the text have been simplified from those presented in this reference source.

40. American Bird Conservancy. 2010. The United States WatchList of Birds of Conservation Concern. Available at http://www.abcbirds.org/abcprograms/science/watchlist/index.html. Accessed January 6, 2012.

41. American Bird Conservancy. 2007. The WatchList 2007. Available at http://www.abcbirds.org/abcprograms/science/watchlist/watchlist.html. Accessed January 6, 2012.

PHOTO 1

Brown pelican (*Pelecanus occidentalis*). This piscivorous (diet consisting mainly of fish) bird, ranging along the south Atlantic and Gulf coasts, was formerly listed as endangered through-out the range. Due to recovery, it was delisted on November 17, 2009 (74 FR 59444–594720) and can now be commonly found in its former habitat. Like other migratory birds, the brown pelican is still protected under the Migratory Bird Treaty Act. There were some calls for relisting following the 2010 British Petroleum Macando Oil spill. (Photographed by Peyton Doub near Crystal River, FL, December 2008.)

PHOTO 2

Bald eagle (*Haliaeetus leucocephalus*). The national symbol of the United States, the bald eagle is a species of unusually sentimental value. This piscivorous bird, ranging over most of North America, was formerly listed as endangered (except in Alaska) and delisted on July 9, 2007 (72 FR 37346–37372) due to recovery. Still protected under the Bald and Golden Eagle Protection Act and the Migratory Bird Treaty Act, the bald eagle is now common near open waters throughout most of the United States. (Photographed by Peyton Doub near Crystal River, FL, December 2008.)

PHOTO 3

American crocodile (*Crocodylus acutus*). This large reptile inhabiting brackish coastal wetlands in south Florida and the Florida Keys is listed as endangered. Although it is the only crocodile to be found in North America, it also ranges in parts of Central and northern South America. The American alligator is listed as "threatened due to similarity of appearance" to this species. (Photographed by Peyton Doub in Everglades National Park, FL, June 2010.)

PHOTO 4

Whooping crane (*Grus Americana*). With fewer than 500 remaining individuals, this large wading bird is listed as endangered. The whooping crane breeds in central Canada and winters on the Texas coast. It is involved in a contentious water use controversy in coastal Texas. Experimental populations are managed by the U.S. Fish and Wildlife Service in Florida and Louisiana. (Photographed by Peyton Doub in the Aransas National Wildlife Refuge, January 2012.)

PHOTO 5
Piping plover (*Charadrius melodus*). This small sandpiper breeds on sandy beaches on the Atlantic, Gulf, and Great Lake coasts. It is listed as endangered (Great Lakes population), threatened elsewhere. The population decline is generally attributed to increased recreational use of beaches. (Photographed by Peyton Doub near Sanibel Island, FL, January 2010.)

PHOTO 6
Florida scrub-jay (*Aphelocoma coerulescens*). This medium-sized passerine bird is, as it name indicates, endemic to the state. Listed as threatened, population declines are generally attributed to loss of oak-dominated upland forest in Florida. (Photographed by Peyton Doub near Marco Island, FL, January 2010.)

4

Related Environmental Statutes and Regulations

4.1 Introduction

The Endangered Species Act does not exist in a vacuum. It is one of dozens of environmental protection laws that have been enacted over the past several decades. The Endangered Species Act is part of a system of federal environmental protection statutes, executive orders, and associated regulations in the United States that is tightly interconnected. Individual environmental protection statutes were initially enacted in response to public concern over specific environmental conditions; in the case of the Endangered Species Act, enactment was a response to public concerns over extinction of several species, such as the passenger pigeon (*Ectopistes migratorius*) and Carolina parakeet (*Conuropsis carolinensis*), and dramatic declines in the populations of others. However, possible species loss has not been the only environmental concern driving environmental regulation in the United States. Losses of historic places prompted enactment of the National Historic Preservation Act in 1966. Concern over the possible environmental effects of federal actions prompted enactment of the National Environmental Policy Act (NEPA) in 1969. Concern over air and water pollution prompted enactment of the Clean Air Act in 1970 and Clean Water Act in 1972. Concern over proper handling and disposal of environmentally hazardous materials led to enactment of the Resource Conservation and Recovery Act (RCRA) in 1976. Concern over chemical contamination prompted enactment in 1980 of the Comprehensive Environmental Response, Compensation, and Liability Act (CERCLA), better known as Superfund. Each of these acts has been strengthened through subsequent legislation, with a notable wave of strengthening occurring in the late 1980s and early 1990s.

For the most part, every element of the environment is subject to some federal environmental regulation. One may think of the environment as comprising air, water, land, and biota (I was a graduate student in the Department of Land, Air, and Water Resources at the University of California at Davis, studying for a master of science degree in plant physiology, earned in 1984.) Air is protected by the Clean Air Act. Water is protected by the Clean Water Act, and sources of drinking water are further protected by the Safe Drinking Water Act. There is no Clean Land Act, but contamination of soils is part of the scope of CERCLA and RCRA. Many sensitive land areas are also addressed in federal regulations, such as wetlands (regulated as part of the Clean Water Act), floodplains (Executive Order 11988), and coastal areas (Coastal Zone Management Act). Protection of biota inhabiting land and water is addressed through a host of specialized regulations, such as the Migratory Bird Treaty Act, the Bald and Golden Eagle Protection Act, and of course the Endangered Species Act.

Just as one cannot effectively protect one component of the environment without considering effects on other components, one cannot effectively administer one of these environmental regulations without effectively administering the others. The Endangered Species Act is no exception. Threatened and endangered species can be killed or otherwise adversely affected by contaminants in the air and water. Threatened and endangered species suffer when their habitats, such as wetlands and coastal areas, are replaced with development or otherwise degraded. Excessive consumption of groundwater that depresses the local water table can dry up springs, streams, and wetlands on which threatened or endangered species might depend. Activities altering the flow patterns or rates in streams and rivers might not only eliminate threatened or endangered aquatic species from certain habitats but also could drown or desiccate threatened or endangered species in associated shorelines and floodplains. Biota, including but not limited to threatened and endangered species, depends on the environmental setting in which it lives.

Table 4.1 lists some of the federal environmental regulations most commonly applicable to land development projects and other projects disturbing naturally vegetated land. It is not a comprehensive list; such a list not only would be long but also would vary with the unique design and circumstances of each project and how one interprets the word *environment*. The importance of comprehensive, individualized evaluation of all potentially applicable environmental compliance requirements as early as possible in the project planning process cannot be overemphasized. Such an evaluation should begin as early as possible in the project conceptualization process, while changes or alternatives are readily practicable, before substantial resources have been committed to any design concept or alternative. Few things can derail a project schedule and budget worse than discovering late in the planning process that an environmental compliance requirement has been overlooked. The Endangered Species Act must be a part of this early comprehensive environmental evaluation, but it should be part of an integrated consideration of all environmental resources and regulatory requirements.

The long list of potentially applicable environmental regulations is one reason behind increasing public apprehension about the complexity of environmental protection and its possible dampening of economic activity. The

TABLE 4.1

Examples of Federal Environmental Acts Potentially Applicable to Land Development Activities

Act	Citation	Regulatory Scope
National Environmental Policy Act	42 U.S.C. 4321 et seq.	Federal actions with potential to significantly impact the human environment
Clean Air Act	42 U.S.C. 7401 et seq.	Emission of air pollutants, including but not limited to six "criteria" pollutants covered by National Ambient Air Quality Standards and greenhouse gases
Clean Water Act Section 402	33 U.S.C. 1342	Discharges of liquid effluent to waters of the United States
Clean Water Act Section 404	33 U.S.C. 1344	Discharges of dredged or fill material (solids) to waters of the United States
Endangered Species Act	16 U.S.C. 1531 et seq.	Actions impacting designated threatened or endangered species or critical habitats
Bald and Golden Eagle Protection Act	16 U.S.C. 668 et seq.	Actions impacting bald and golden eagles
Migratory Bird Treaty Act	16 U.S.C. 703 et seq.	Actions impacting migratory birds native to the United States
National Historic Preservation Act	16 U.S.C. 470 et seq.	Actions impacting historic or archaeological resources eligible for the National Register of Historic Places
Farmland Protection Policy Act	7 U.S.C. 4201 et seq.	Federal actions causing the loss of prime or unique farmland
Resource Conservation and Recovery Act	42 U.S.C. 6901 et seq.	Generation, handling, and disposal of hazardous materials "cradle to grave"
Comprehensive Environmental Response, Compensation, and Liability Act	42 U.S.C. 9601 et seq.	Identification and cleanup of environmentally contaminated sites
Safe Drinking Water Act	21 U.S.C. 301 et seq.	Actions affecting drinking water sources
Coastal Zone Management Act	16 U.S.C. 1451 et seq.	Actions proposed for areas close to oceans, estuaries, and the Great Lakes
National Wild and Scenic Rivers Act	16 U.S.C. 1271–1287	Federal actions impacting rivers designated as wild and scenic
Native American Graves Protection and Repatriation Act	25 U.S.C 3001 et seq.	Actions impacting burial grounds associated with Native American (including Eskimo and Native Hawaiian) tribes

apprehension is partially justified; the complex web of interrelationships between various components of the environment does not lend itself to a simplistic planning process. Furthermore, each component in the array of environmental statutes is the product of its own constituency and its own history. Each arose out of the concern by some sector of the American public that worried over the loss or degradation of some environmental feature. This worry culminated in passage of a law, development of regulations, and establishment of a bureaucracy of agency employees to administer the regulations and a parallel establishment through the free market of a field of consultants and consulting companies vying to be hired out to assist the bureaucrats and project proponents seeking authorizations from those bureaucrats. Different cadres of bureaucrats, consultants, activists, and other stakeholders in each environmental statute all guard their turf; a waning breadth of perceived applicability of the statute could mean loss of careers. A project proponent who finds a consultant able to identify the full breadth of environmental compliance requirements without overstating the interests of any one specialized constituency is fortunate indeed.

The following sections examine several of the federal environmental protection statutes (and executive orders) most intimately intertwined with the Endangered Species Act. They do not address every statute connected in some way to the Endangered Species Act. Nor do the sections provide comprehensive introductions to the statutes they cover. The purpose of the sections is to introduce project managers and environmental consultants to the breadth of environmental regulation that is closely intertwined with the Endangered Species Act. Project proponents must also consider environmental regulations imposed at the state level in the state where their projects are proposed. State endangered species regulations and other state environmental regulations intimately associated with endangered species protection are discussed in Chapter 7.

4.2 The National Environmental Policy Act

NEPA is a comprehensive environmental planning process for federal actions that overlaps considerably with the Endangered Species Act. NEPA was enacted less than four years prior to the Endangered Species Act, and each derives from many of the same environmental concerns expressed by the American public in the years after World War II. The scope of NEPA is considerably broader than that of the Endangered Species Act; in fact, it extends to the entirety of the "human environment," a term that is inclusive not only of rare animal and plant species but also topics as diverse as air and water quality, waste management, recreation, aesthetics, historical and archaeological resources, noise, traffic, and even effects on the local economy. While most environmental statutes call for consideration of specific environmental resources, NEPA calls for an integrated evaluation of possible environmental impacts. Furthermore, while most environmental statutes focus on protecting resources and require permits or approvals from agencies with specialized knowledge of the protected resources, NEPA focuses more on informed decision making among possible project alternatives and disclosure of those decisions and possible environmental impacts to the public.

The scope of NEPA also extends to ecological issues not within the purview of the Endangered Species Act, including effects on animal and plant species and habitats not listed as threatened or endangered or as critical habitat. Table 4.2 provides a more inclusive but not comprehensive list of

TABLE 4.2

Topics Commonly Covered in Environmental Impact Statements

Topic	Typical Considerations
Land use	Land availability, compatibility with existing and proposed uses of nearby property, compliance with local zoning ordinances, compatibility with regional comprehensive land use plans. Specialized land planning topics such as impacts to the coastal zone, prime and unique farmlands, and wild and scenic rivers.
Water resources	Use and availability of surface and groundwater resources. Wetlands are commonly addressed in this section or in the biological resources section.
Biological resources	Wildlife, vegetation, hunting and fishing, and federal and state-listed threatened and endangered species and critical habitats. Wetlands are commonly addressed in this section or in the water resources section.
Socioeconomics	Population and demographic issues, demands on public services such as fire and police, availability of housing and lodging services, and opportunities for education and employment.
Cultural resources	Historical and archaeological resources eligible for inclusion on the National Register of Historic Places.
Air quality	Emissions of air pollutants, including but not limited to "criteria" pollutants regulated by National Ambient Air Quality Standards and greenhouse gases.
Traffic	Traffic congestion on existing and planned highways. For some projects, might consider air, rail, or water transportation.
Noise	Generation of sound that could disturb residential and other land uses. Sometimes addressed in aesthetics section.
Aesthetics/visual resources	Visual compatibility with existing and future land uses.
Soils and geology	Soil erosion and sedimentation of waterways. Compatibility with regional geological conditions. Sometimes considers effects on availability of mineral resources.
Safety	Effects on human safety and well-being. For some actions, may consider effects of radiation.

environmental topics that can potentially fall under the scope of NEPA. Indeed, any such list cannot be comprehensive, as NEPA provides for an individualized "scoping" process for individual projects. One can in fact state that the potential breadth of NEPA is limited only by the potential diversity of possible federal projects.

The statutory language of NEPA reflects the desire to provide for comprehensive consideration of the environmental ramifications of federal actions at a time when a "space age" society contemplated ever more ambitious efforts to subjugate and manipulate the environment. The stated purpose of NEPA is

> to declare a national policy which will encourage productive and enjoyable harmony between man and his environment, to promote efforts which will prevent or eliminate damage to the environment and biosphere and stimulate the health and welfare of man, to enrich the understanding of the ecological systems and natural resources important to the Nation, and to establish a Council on Environmental Quality.[1]

Note the inclusion of "ecological systems" in the purpose. The Council on Environmental Quality (CEQ) established under NEPA was rapidly eclipsed in its role by the newly established Environmental Protection Agency (EPA) but remains to the present day as a small advisory body to the president. CEQ still, however, issues regulatory updates and general guidance on NEPA and helps to mediate disputes regarding NEPA compliance.

The best known, but certainly not only, element of NEPA is the requirement for federal agencies to issue an environmental impact statement (EIS) for "legislation and other major Federal actions significantly affecting the human environment." The term EIS is not used in the statutory language of NEPA; the act specifically calls for a "detailed statement" addressing:

- The environmental impact of the proposed action,
- Any adverse environmental effects which cannot be avoided should the proposal be implemented,
- Alternatives to the proposed action,
- The relationship between local short-term uses of man's environment and the maintenance and enhancement of long-term productivity, and
- Any irreversible and irretrievable commitments of resources which would be involved in the proposed action should it be implemented.[2]

NEPA is less prescriptive than the Endangered Species Act. It does not specifically limit what federal agencies can do. It does not call for any permit, such as a take permit. It does not prevent federal agencies from taking actions that adversely affect the environment, even if those effects are significant. What NEPA does do is call for disclosure of possible environmental impacts prior to implementation. Agencies can affect the environment in any way necessary, just as long as the agency attempts to evaluate and describe the impacts to the public and consider alternatives prior to deciding to embark on an action.

Note the requirement to evaluate alternatives to the proposed action. CEQ refers to the consideration of alternatives as the "heart of the EIS," and most EISs consist of detailed discussions of possible environmental effects from multiple alternatives. The expectation is that by considering alternatives to a contemplated action, agencies might discover and implement alternatives that meet their objectives equally well but with fewer impacts on the environment. The logic is hardly unique; who among us does not consciously or subconsciously contemplate possible alternatives prior to making major decisions—alternative cars or houses, alternative vacation plans, alternative career paths, and so on.

Like the Endangered Species Act, the details for implementation of NEPA are contained not in the statutory language but in regulations. CEQ issued regulations in 1978 that are commonly referred to as the "CEQ guidelines"[3] that establish a general road map for how federal agencies should comply with NEPA. But, each agency establishes its own specific regulations for NEPA compliance. Table 4.3 lists the NEPA regulations for several key agencies. The various agency-specific regulations share a lot of commonalities; most central are a requirement to evaluate potential environmental impacts

TABLE 4.3

NEPA Regulations for Select Federal Agencies

Agency	Bureau	NEPA Regulations
Department of Agriculture	Forest Service	36 C.F.R. 220
	Natural Resources Conservation Service	7 C.F.R. 650
Department of the Interior	Fish and Wildlife Service	43 C.F.R. 46
	National Park Service	43 C.F.R. 46
	Bureau of Land Management	43 C.F.R. 46
Department of Commerce	National Marine Fisheries Service	48 F.R. 14734
Department of Defense	Army	32 C.F.R. 651
	Navy	32 C.F.R. 775
	Air Force	32 C.F.R. 989
	Army Corps of Engineers	33 C.F.R. 230
Department of Energy		10 C.F.R. 1021
Department of Transportation	Federal Aviation Administration	FAA Order 5050.4B
	Federal Highway Administration	23 C.F.R. 771
Environmental Protection Agency		40 C.F.R. 6
Nuclear Regulatory Commission		10 C.F.R. 51
Federal Energy Regulatory Commission		18 C.F.R. 380

for a range of reasonable alternatives. All are, however, tailored to the unique aspects of how individual agencies operate.

Although the most public face of how most agencies comply with NEPA is the EIS, few federal actions are actually preceded by an EIS. NEPA requires agencies to prepare a "detailed statement" only for "major federal actions significantly affecting the human environment." Not every action contemplated by an agency is "major" or capable of "significantly" affecting the environment. The CEQ regulations, and those of most agencies, allow agencies to prepare short "environmental assessments" (EAs) for small actions requiring individualized review to determine whether they could potentially result in significant environmental impacts. The regulations further encourage agencies to establish categorical exclusions (CEs) under which specific categories of minor actions are determined in advance not to require an EIS without individualized review. Many other environmental regulations, most notably the Clean Water Act with its "general permits," have similar provisions for expedited or simplified review of minor or routine actions; this flexibility is intended to allow the regulations to function in a practical manner without excessively encumbering the economy. There are no general take permits under the Endangered Species Act, but the no surprises goal of the habitat conservation plans (discussed in Chapter 6 of this book) allowed under the act offers a somewhat parallel flexibility.

The CEQ regulations specifically encourage federal agencies to integrate into the NEPA process compliance efforts under several related environmental requirements, including those of the Endangered Species Act. They state:

> To the fullest extent possible, agencies shall prepare draft environmental impact analyses and related surveys and studies required by the Fish and Wildlife Coordination Act (16 USC 661 et seq.), the National Historic Preservation Act of 1966 (16 USC 470 et seq.), the Endangered Species Act of 1973 (16 USC 1531 et seq.), and other environmental review laws and executive orders (40 CFR 1502.25).[4]

Most agencies informally confer with the U.S. Fish and Wildlife Service (FWS) and the National Marine Fisheries Services (the Services) in the early stages of planning an EIS or EA. This effort falls within the purview of what the Services term "informal consultation," discussed further in Chapter 5. At a minimum, action agencies or their consultants should seek relevant data from the applicable state natural heritage program and ideally they actually describe the proposed action and possible alternatives with representatives of the Services. A key element of EISs and some more complex or controversial EAs is scoping, a process in which the action agency presents the action and invites comments from the public and other interested federal, state, and local agencies. Scoping serves to identify potential issues and concerns and helps tailor the content of the EIS or other NEPA document. The Services

may choose to weigh in during scoping on possible issues under their purview, including threatened and endangered species.

Most EISs, but few EAs, are issued first as drafts for public comment
(draft EIS), followed by a final version (final EIS) that responds to comments. The Services will commonly issue comment letters on draft EISs
that involve impacts to threatened or endangered species, critical habitats,
or other resources under the Endangered Species Act or other regulations
under the Services' purview. The comment letters are not formal consultations under Section 7 of the Endangered Species Act, although they might
express a need for the action agency to complete a biological assessment
and initiate formal consultation. In other cases, the comment letter might
indicate that further involvement of the Services is not necessary (sometimes termed "no further action" letters). Effort by the Services to comment
on a draft EIS, as when commenting at the scoping stage, also falls within
the purview of "informal consultation." As noted in Chapter 5, the conduct
of informal consultation does not necessarily presage formal consultation;
indeed, one objective of informal consultation is to resolve problems and
reduce impacts enough early during the planning process to obviate the
need for formal consultation.

The requirements of NEPA, like those of Section 7 of the Endangered
Species Act, extend only to actions proposed by federal agencies and not,
at least in a direct sense, to those proposed by state or local agencies or by
private parties. However, nonfederal actions can and often do become "federalized" through the need for federal permits or authorizations (or funding). In particular, the need for a federal permit from the U.S. Army Corps
of Engineers under the Clean Water Act to authorize work in wetlands or
surface water bodies (see Section 4.3 for further discussion) federalizes many
land development projects by private developers. The need for federal permits, especially those under the Clean Water Act, is also what brings many
privately proposed land development projects into the realm of Section 7
consultation under the Endangered Species Act. Many consultants and
consulting firms specialize in obtaining federal, state, and local permits for
private land development projects, and it is this federalization process that
brings those consultants into the processes of both NEPA and Section 7.

The foregoing is but a brief introduction to NEPA with an emphasis on
how it interplays with the Endangered Species Act. Several books are available to provide comprehensive overviews of NEPA at a level comparable to
how this book addresses the Endangered Species Act. To close, a few key
points need to be stated regarding the interrelationship of NEPA and the
Endangered Species Act:

- The scope of NEPA, including the scope of biological impact assessment under NEPA, is not limited to listed threatened or endangered
  species or critical habitat; indeed, the scope of NEPA review can

extend to "common" species and habitats as well. NEPA scoping can also involve species designated as proposed, candidate, or special concern under the Endangered Species Act and state-designated species and habitats.

- The fact that an action qualifies under a CE or FONSI (finding of no significant impact) does not necessarily indicate that formal Section 7 consultation or permits under the Endangered Species Act are not necessary. Likewise, the fact that an action does not require formal Section 7 consultation or Endangered Species Act permits does not necessarily indicate that the action could qualify under a CE or FONSI.

- The CEQ encourages agencies to coordinate NEPA and Endangered Species Act compliance as closely as practicable, but the two acts have separate requirements that must in the end be met. But, agencies do not always coordinate the two compliance efforts in a predictable pattern.

- Action agencies and their consultants undertaking NEPA documentation efforts should seek out data on threatened and endangered species and critical habitats as early as possible. For an EIS, the effort should precede scoping. The data should be used when initially conceptualizing the action; altering actions and switching to alternatives is usually much easier and less costly early in the process. Species and habitat data should be used when performing formal siting or routing reports for large projects, but equally important is use of that data in less-formal conceptual planning efforts for smaller projects, including those addressed in EAs or CEs.

- The Services should be invited by action agencies to participate in NEPA scoping and commenting processes, and informal input should be sought from the Services even for projects proceeding under EAs or CEs. If the Services receive enough information in these early stages to determine confidently that the proposed action would not adversely affect listed resources, they will issue a "no further action" letter indicating that formal Section 7 consultation is not required. The formal consultation process, if it is required, will rarely raise issues not already raised informally. Only when the Services are not approached until formal consultation is there a significant possibility of unforeseen issues resulting in substantial delays.

- The CEQ regulations and most agency-specific NEPA regulations direct preparers of EISs and EAs to list environmental permit requirements and persons and agencies consulted during the NEPA process; the Endangered Species Act and the Services are included in these listings for most actions affecting natural habitats.

4.3 The Clean Water Act

The Endangered Species Act and the Clean Water Act are interrelated in ways that might not seem obvious to many outside the environmental profession. What is obvious is that clean water benefits threatened and endangered species that live in and drink from water sources. Less obvious but no less logical is that clean water benefits food chains made up of species, including but not limited to threatened or endangered species, that feed on other species, all dependent on clean water. That part of the Clean Water Act most directly addressing the prevention of water pollution is Section 402, more commonly referred to as the National Pollution Discharge Elimination System (NPDES). Parties discharging liquids into streams, rivers, or other surface water bodies in the United States must do so under an NPDES permit. An NPDES permit establishes limits with respect to various physical and chemical parameters of the discharged liquid that cannot be exceeded along with a requirement for periodic (usually monthly) monitoring of the discharges. The points at which discharges are released to regulated bodies of water are termed outfalls. The monthly monitoring reports for each outfall are termed discharge monitoring reports (DMRs).

Examples of parameters typically addressed in NPDES permits and DMRs include temperature, pH (acidity), total dissolved solutes, conductivity, chemical nutrients such as nitrogen and phosphorus, metals such as iron and zinc, volatile and semivolatile organic compounds, and pesticides. The limits in NPDES permits are set in accordance with water quality criteria established by federal and state regulations. Federal water quality regulations are established by the EPA following publication in the *Federal Register* and receipt of public comments. EPA must republish any proposed changes to federal water quality criteria and respond to a new round of public comments. Scientists with EPA cannot alter federal water quality criteria without receiving input from the public. The Clean Water Act also allows states to establish state water quality criteria, and most states have done so. To be enforceable under the Clean Water Act, state water quality criteria must be at least as strict as corresponding federal criteria for corresponding parameters. The federal criteria therefore establish minimum thresholds across the nation. States may also establish state water quality criteria for parameters lacking federal criteria, and federal criteria apply for parameters lacking state criteria.

Although the NPDES is administered federally by EPA, many states have requested and received "delegation" of NPDES so that NPDES permits are issued by agencies of those states. Many states "personalize" the name of their NPDES permits; for example, Virginia refers to theirs as "Virginia Pollutant Discharge Elimination System (VPDES) Permits." DMRs under those permits are submitted to the state agency.

Traditionally, NPDES permits were issued for "end-of-pipe" outfalls discharging liquid effluent to discrete points on surface water bodies. Pollution from such outfalls is commonly referred to as "point source" water pollution. These outfalls are typically associated with factories, power plants, or municipal sewage treatment plants. Although some of the specific water quality criteria have been, and continue to be, controversial, the need to receive and comply with such permits is easy to understand and generally not especially controversial. More controversial has been the push since around 1990 to require NPDES permits for "non-point source" liquid discharges. These discharges typically originate from precipitation falling on developed areas and flowing (running off) into streams and other surface water bodies. This runoff is commonly termed *stormwater runoff.* Unlike point source discharges, stormwater runoff does not enter water bodies by way of pipes releasing at discrete locations. The locations at which it enters are less obvious and may be numerous, although knowledgable hydrologists can determine the locations. While requirements to permit and monitor point source outfalls typically fall on industrial and municipal parties, stormwater runoff can also result from residential and commercial construction sites and agricultural operations, including small farms. Many of these entities lack the resources to comply with environmental regulations and may be unaware that they are subject to the requirements.

Considerably more controversial than NPDES, and more tightly intertwined with the Endangered Species Act, is Section 404 of the Clean Water Act. Section 404 requires permits for the "discharge of dredged or fill material" into "waters of the United States." A simple, but not entirely accurate, analogy is that while Section 402 (NPDES) covers discharges of liquids into our nation's waterways, Section 404 covers discharges of solids, usually soils or sands. Such a requirement may not seem more controversial than NPDES until one considers how the courts have interpreted "waters of the United States." In *United States v. Riverside Bayview Homes, Inc.,*[5] the Supreme Court ruled that the requirements of Section 404 extended not only to traditional navigable waters such as oceans, estuaries, and large lakes and rivers, but also to wetlands adjacent to those waters. The U.S. Army Corps of Engineers has published detailed regulations on the scope of "waters of the United States" in 33 C.F.R. 328.3. Waters of the United States include most surface water features such as streams, rivers, lakes, estuaries, and seas, but also include less-intuitive features. such as "mudflats, sandflats, wetlands, sloughs, prairie potholes, wet meadows, playa lakes, or natural ponds."[6] Section 404 is commonly referred to as the federal wetland permitting program, although its scope is considerably broader than just wetlands.

The spatial extent of rivers, lakes, and surface water bodies is defined under the Clean Water Act as the logical and easy-to-define ordinary highwater mark, which is defined as

that line on the shore established by the fluctuations of water and indicated by physical characteristics such as clear, natural line impressed on
the bank, shelving, changes in the character of soil, destruction of terrestrial vegetation, the presence of litter and debris, or other appropriate
means that consider the characteristics of the surrounding areas.[7]

The inclusion of wetlands in the act's scope led to heated controversy
regarding where the regulated wetlands end and the unregulated nonwetlands (commonly termed uplands) begin. In short, while everyone knows
the difference between land and river, stream, or lake, not everyone intuitively knows whether an area is a wetland. The central wettest part of most
large swamps and marshes is usually shallowly inundated or at least saturated (soggy), but the edges are often dry. The Clean Water Act establishes a
regulatory definition for wetlands that is of little clarifying value. It defines
wetlands as

areas that are inundated or saturated by surface or ground water at a
frequency and duration sufficient to support, and that under normal
circumstances do support, a prevalence of vegetation typically adapted
for life in saturated soil conditions. Wetlands generally include swamps,
marshes, bogs, and similar areas.[8]

This is a highly technical definition whose use requires multidisciplinary
scientific expertise. Notice the combined references to hydrological conditions (saturation and inundation) and vegetation. The U.S. Army Corps
of Engineers recognized that the regulatory definition did not provide an
adequate technical basis for identifying where wetlands end and dry land
begins. They published in 1987 a technical manual for delineating wetlands
termed the *Corps of Engineers Wetlands Delineation Manual.*[9] The procedure
identifies wetlands based on indicators of three parameters: vegetation
(hydrophytic vegetation), soils (hydric soils), and hydrology (wetland hydrology). This publication launched the careers of countless technical consultants (I am among them) specializing in application of the manual's technical
procedures and writing reports termed "wetland delineation reports." The
U.S. Army Corps of Engineers has since published multiple regional supple-
ments to the 1987 technical manual that provide regionally tailored guidance
for delineating wetlands.

Most wetland scientists recognize the role wetlands play in providing
habitat for threatened and endangered species. The process of describing
the specific roles wetlands play in the environment goes under many names
but is most commonly referred to as "wetland functional assessment."
Functional assessment methodologies can be descriptive or semiquantitative. The oldest widely recognized federal technique, the wetland evaluation
technique (WET), uses the responses to a questionnaire to predict whether a
wetland may provide specific functions and values. Functions are physical,

chemical, or biological activities that directly benefit society or the environment. Values are indirect social benefits such as aesthetic qualities or availability for recreation.[10] The known occurrence of threatened or endangered species was identified by WET as one of several "red flags" that automatically elevated the evaluated wetland to a high prediction of environmental quality, even in the absence of other predictors of quality.

The New England District of the corps developed a structured descriptive approach, called the highway methodology.[11] Specific values and functions considered by the highway methodology include the following:

- groundwater recharge and discharge (function);
- flood flow alteration (function);
- fish and shellfish habitat (function);
- sediment, toxicant, and pathogen retention (function);
- nutrient removal, retention, and transformation (function);
- production export (function);
- sediment and shoreline stabilization (function);
- wildlife habitat (function);
- recreation (value);
- education and scientific value (value);
- uniqueness and heritage (value);
- visual quality and aesthetics (value); and
- threatened or endangered species habitat (value).

Several semiquantitative functional assessment methods involve the calculation of scores based on geographic, physical, and biological properties. Some methods, such as the corps' hydrogeomorphic approach,[12] compare scores against corresponding "reference" wetlands. Other methods, such as the Florida Uniform Mitigation Assessment method (Chapter 62-345 of the Florida Administrative Code),[13] generate scores based on observed conditions. Environmental Concern, Incorporated, has summarized 40 separate functional assessment methodologies.[14] When selecting a methodology, wetland scientists have to consider regional suitability and ease of use.

4.4 The Fish and Wildlife Coordination Act

The Fish and Wildlife Coordination Act is one of the oldest conservation acts, dating from the 1930s, and is the oldest of the "consultation" requirements in environmental planning. The biological breadth of the act is not limited to designated rare species but to the entirety of fish and wildlife. But, threatened and endangered species of fish and wildlife are certainly within the act's scope of consideration. Like Section 7 of the Endangered Species Act, it applies only to federal actions, although those actions may involve decisions to permit, authorize, or fund privately sponsored development activities. It is much less prominent and far less controversial than the Endangered Species Act. Many environmental consultants engage in "informal consultation" with the FWS regarding Section 7 (see Section 5.4 of this book) and receive a response that no further action is required under Section 7 or the Fish and Wildlife Coordination Act, almost a furtive "silent" informal consultation exercise.

The Fish and Wildlife Coordination Act is directed at federal water resource development projects. It grew out of public concern over dam building and other activities affecting rivers and other surface water features providing habitat for fish and wildlife in the western United States. The act's stated purpose is

> recognizing the vital contribution of our wildlife resources to the Nation, the increasing public interest and significance thereof due to expansion of our national economy and other factors, and to provide that wildlife conservation shall receive equal consideration and be coordinated with other features of water-resource development programs through the effectual and harmonious planning, development, maintenance, and coordination of wildlife conservation and rehabilitation.[15]

Placing the interests of fish and wildlife on the same plane as agricultural, mining, and other development interests was a revolutionary concept at the time the act was initially implemented. The act is often associated with dams and other river development projects proposed by the U.S. Army Corps of Engineers, but the actions of other federal agencies are covered.

The Fish and Wildlife Coordination Act establishes a separate consultation requirement, independent of that required under the Endangered Species Act, for federal projects affecting water resources. But, it need not be, and usually should not be, a separate effort. Both at the informal and the formal stages of Section 7 (discussed in detail in Section 5.4 of this book), action agencies and their consultants should simultaneously address the Fish and Wildlife Coordination Act. The Services usually communicate to agencies in tandem regarding both acts. The interpretation of which projects are covered under the act is not always clear. The Services will, during the informal communications with action agencies under Section 7, also indicate what, if any, requirements must be performed under the Fish and Wildlife Coordination Act. Agencies and their consultants must, however, ensure that the lesser-known act is not overlooked.

4.5 The Bald and Golden Eagle Protection Act

Few federal environmental protection statutes target specific species. The bald eagle (*Haliaeetus leucocephalus*), our nation's national symbol, and the very similar golden eagle (*Aquila chrysaetos*) are two exceptions. Few species stir our national passion like the bald eagle, a piscivorous (fish-eating) raptor (predatory bird) that adorns our money, most federal stationery, the seals of many federal agencies, and many other items associated with the federal government. The golden eagle is no less majestic a piscivorous raptor, although less widely recognized than the bald eagle. The Bald and Golden Eagle Protection Act, sometimes casually referred to as the Eagle Act, prohibits actions that

> knowingly, or with wanton disregard for the consequences of this act take, possess, sell, purchase, barter, offer to sell, purchase or barter, transport, export or import, at any time or in any manner any bald eagle commonly known as the American eagle or any golden eagle, alive or dead, or any part, nest, or egg thereof of the foregoing eagles.[16]

"American eagle" is another common name for the bald eagle. Until June 2007, the bald eagle was also protected in most geographic areas under the Endangered Species Act, first as endangered, then since 1995 as threatened.[17] The protections offered under the Eagle Act were in addition to those under the Endangered Species Act. Until delisting, the bald eagle was addressed in countless Section 7 consultations and associated biological assessments and biological opinions, which usually mentioned the fact that the species was also protected under the Eagle Act but otherwise addressed this no further. Those concerned over possible unregulated impacts to the bald eagle were reassured that the species would continue to be protected under the Eagle Act (as well as the broader Migratory Bird Treaty Act, discussed further in this chapter), and that the FWS would continue to monitor the species for a minimum of five years for possible relisting.[18] Immediately on the bald eagle's delisting, the once-obscure Eagle Act took on new meaning as a bulwark against unregulated impacts on our national symbol.

Notice the similarity in what the Eagle Act prohibits for eagles and what the Endangered Species Act prohibits for threatened or endangered species. At first glance, the Eagle Act offers everything to protect the bald eagle that it received under the Endangered Species Act. But the word *take* has been subject to disparate interpretation. Traditionally, the Eagle Act was interpreted as prohibiting shooting or otherwise directly killing or molesting eagles or commerce in the species. Note the words *knowingly* and *with wanton disregard for the consequences of this act* in the regulatory prohibitions established in the act. As for the Endangered Species Act, prohibitions against "incidental" take, specifically habitat disturbance caused by land development, were not contemplated over most of the Eagle Act's history. Once prohibitions on incidental take became well established under the Endangered Species Act, environmental activists noticed that habitat protections afforded in connection with preventing incidental take of the bald eagle concurrently protected a lot of environmentally sensitive coastal, shoreline, and riparian (streamside) habitat. Even those activists who grudgingly acknowledged that the bald eagle had become much more common than in the early years of the Endangered Species Act were concerned that official "recovery" and delisting would free up some of these protected habitats from federal regulation. Around a month prior to delisting the bald eagle, the FWS revised the definition of "disturb" under the Eagle Act to the following:

> To agitate or bother a bald or golden eagle to a degree that causes, or is likely to cause, based on the best scientific information available, (1) injury to an eagle, (2) a decrease in its productivity, by substantially interfering with normal breeding, feeding, or sheltering behavior, or (3) nest abandonment, by substantially interfering with normal breeding, feeding, or sheltering behavior.[19]

This definition effectively extends protection under the Eagle Act to habitat known to be used by the bald eagle for nesting or foraging, essentially prohibiting disturbance of such habitat incidental to most forms of land development or industrial activity.

Both eagle species are widely distributed across most of the United States. Federal action agencies and their consultants should be sure to include the Eagle Act in their informal communications with the Services regarding the Endangered Species Act and the Fish and Wildlife Coordination Act. The Services welcome an integrated early communication effort regarding the totality of the various wildlife conservation regulatory requirements. Consultants for privately sponsored land development projects should also not overlook the Eagle Act, not only because of federalization through the need for federal permits or authorization but also because the requirement for take permits under the Eagle Act is not limited to federal actions.

The take permit requirements established under the Eagle Act are highly parallel to those under the Endangered Species Act. However, the Eagle Act does not establish a formal consultation process parallel to Section 7. The FWS does, however, have a more than 20-year history of informal and formal consulting with federal agencies on the bald eagle from when it was still listed as endangered and then threatened. The Service will provide project planners with information needed to make early adjustments to proposed actions to avoid or reduce eagle impacts. If impacts are unavoidable, the Service will direct project proponents through the process of obtaining eagle take permits, including development of appropriate mitigation measures. Although the Eagle Act does not establish a formal consultation requirement, informal communication with the FWS in a manner comparable to informal consultation under the Endangered Species Act (Section 5.4 of this book) is an important element of sound environmental planning.

In response to uncertainty over how bald eagle impacts would be regulated following delisting of the bald eagle, the FWS published detailed national guidelines outlining measures to minimize possible disturbance of the bald eagle.[20] Although the guidelines are not regulations, they do provide project proponents with insight into how the FWS focuses enforcement activities under the Eagle Act. Specific recommendations are provided for each of eight categories of activity, including projects with minor potential impacts (Category A), projects with major potential impacts (Category B), timber harvest and forestry (Category C), off-road vehicle use (Category D), motorized watercraft use (Category E), nonmotorized recreation and entry (Category F), helicopter and aircraft use (Category G), and noise generation (Category H). The recommendations generally call for limiting activity during the bald eagle breeding season within specified distances of nest sites, ranging from 330 feet for activities with generally localized impacts to as much as 0.5 mile for generation of loud noises. The recommended distance for most large development projects, including but not limited to developments of over 0.5 acre, mining, and oil and gas drilling, is between 330 and 660 feet, depending on visibility from the nest and whether similar activities are already present in the localized area. Less-restrictive recommendations are provided for activities near bald eagle foraging areas and communal roost sites.

The guidelines provide additional recommendations to help minimize disturbance to the bald eagle. They are indicative of some of the specific types of effects that many development projects can have on this species. They include

1. Protect and preserve potential roost and nest sites by retaining mature trees and old growth stands, particularly within ½ mile from water;
2. Where nests are blown from trees during storms or are otherwise destroyed by the elements, continue to protect the site in the absence of the nest for up to three (3) complete breeding seasons. Many eagles will rebuild the nest and reoccupy the site;
3. To avoid collisions, site wind turbines, communication towers, and high voltage transmission power lines away from nests, foraging areas, and communal roost sites;
4. Employ industry-accepted best management practices to prevent birds from colliding with or being electrocuted by utility lines, towers, and poles. If possible, bury utility lines in important eagle areas;
5. Where bald eagles are likely to nest in human-made structures (e.g., cell phone towers) and such use could impede operation or maintenance of the structures or jeopardize the safety of the

eagles, equip the structures with either (1) devices engineered to discourage bald eagles from building nests, or (2) nesting platforms that will safely accommodate bald eagle nests without interfering with structure performance;

6. Immediately cover carcasses of euthanized animals at landfills to protect eagles from being poisoned.

7. Do not intentionally feed bald eagles. Artificially feeding bald eagles can disrupt their essential behavioral patterns and put them at increased risk from power lines, collision with windows and cars, and other mortality factors.

8. Use pesticides, herbicides, fertilizers, and other chemicals only in accordance with Federal and state laws.

9. Monitor and minimize dispersal of contaminants associated with hazardous waste sites (legal or illegal), permitted releases, and runoff from agricultural areas, especially within watersheds where eagles have shown poor reproduction or where bioaccumulating contaminants have been documented. These factors present a risk of contamination to eagles and their food sources.[21]

The FWS implemented formal regulations for incidental take of bald and golden eagles under the act in 2009.[22] These regulations establish a permitting process and conditions for eagle take "that is associated with, but not the purpose of, an activity," that is, incidental take, for individual instances of take that cannot be practicably avoided or for programmatic take when take is unavoidable even though advanced conservation measures are being implemented. Key conditions are a requirement to take reasonable steps to avoid, minimize, and mitigate for the effects of the take (a common theme shared with many other environmental regulations, such as NEPA and Section 404 of the Clean Water Act), to monitor eagle activities in the action area throughout the permitted work plus up to three years after completion, and to submit annual reports throughout the monitoring period. The FWS issued a draft EA for the first instance of permitted take applied for under the new regulations in December 2011, which was for take of one to two golden eagles over the anticipated 20- to 30-year operating life of a proposed wind energy project in Oregon.[23]

4.6 The Migratory Bird Treaty Act

Many threatened and endangered species, as well as the bald and golden eagles protected under the Eagle Act, are also migratory birds. Well-known examples of threatened or endangered species that are migratory birds include the whooping crane (*Grus americana*), piping plover (*Charadrius melodus*), and wood stork (*Mycteria americana*). Birds stir passions in ways that many other groups of species, even threatened or endangered species, do not. This fact may reflect the millions of Americans who consider themselves bird-watchers (or, as many bird-watchers like to refer to themselves as, birders). Concern for birds, and the popularity of bird-watching as a hobby, may ultimately derive from the ubiquitous presence of many bird species even in urban and suburban settings, the colorful nature and diversity of birds, the way birds respond to feeders, the challenge of compiling observations in life lists of species, and the allure of travel to encounter ever more species. Most bird species in North America, including most North American species protected under the Endangered Species Act, are migratory. Extinction of several North American migratory bird species such as the passenger pigeon is one factor contributing to establishment of the Endangered Species Act.

Migratory birds are birds that move substantial distances from their breeding (summer) habitat to their nonbreeding (winter) habitat. They are sensitive to impacts that resident birds remaining in one geographic locale year-round are not. Habitat damage at only one of the breeding or wintering grounds can be detrimental to individuals present in either setting. Attempts to protect or conserve a migratory bird species in one locality can be futile if damages continue at the other. Migratory birds are also sensitive to effects on intervening areas along their migration routes. Areas crossed by migration routes for several migratory bird species are commonly termed flyways.

The Migratory Bird Treaty Act makes it unlawful to

> pursue, hunt, take, capture, kill, attempt to take, capture, or kill, possess, offer for sale, sell, offer to barter, barter, offer to purchase, purchase, deliver for shipment, ship, export, import, cause to be shipped, exported, or imported, deliver for transportation, transport or cause to be transported, carry or cause to be carried, or receive for shipment, transportation, carriage, or export, any migratory bird, any part, nest, or eggs of any such bird, or any product, whether or not manufactured, which consists, or is composed in whole or part, of any such bird or any part, nest, or egg thereof.

The act is limited in scope to bird species identified in specific treaties between the United States and Great Britain, Mexico, Japan, and the former Soviet Union. The inclusion of Mexico is particularly important as many migratory birds that summer (breed) in the United States winter in Mexico.

The protections afforded under the act are limited to migratory birds that are native to the United States. The act specifically excludes from its scope bird species "occurring in the United States or its territories solely as a result of intentional or unintentional human-assisted introduction" unless it was formerly native to the United States prior to 1918. The act may be generally thought of as a native bird protection act. It affords protection for many species, rare and common, not listed as threatened or endangered under the Endangered Species Act. Species as common as the American robin and song sparrow are protected. Introduced bird species such as the house sparrow and starling are not protected. Protections for threatened or endangered species are in addition to those afforded under the Endangered Species Act.

The FWS administers permits "for the taking, possession, transportation, sale, purchase, barter, importation, exportation, and banding or marking of migratory birds" under regulations established in 50 C.F.R. 21. Permits are specifically available under 50 C.F.R. 21 Subpart C for

- importing or exporting migratory birds or their components (e.g., nests or eggs) (50 C.F.R. 21.21);
- capturing migratory birds for banding or marking, typically for purposes of scientific research (50 C.F.R. 21.22);
- scientific collection of migratory bird specimens (50 C.F.R. 21.23);
- taxidermy of migratory birds (50 C.F.R. 21.24);
- sale or trade in waterfowl (e.g., most ducks) (50 C.F.R. 21.25);
- activities to control or manage Canada geese, a migratory species that is a pest in some settings (50 C.F.R. 21.26);
- falconry (50 C.F.R. 21.29);
- possession or trade in raptors (birds of prey) (50 C.F.R. 21.30);
- wildlife rehabilitation (50 C.F.R. 21.31); or
- other purposes (sort of a catchall provision) (50 C.F.R. 21.27).

The FWS also issues "Depredation Permits" for activities that farmers or other property owners might take to prevent or limit damage from migratory birds to crops or other property.[24] The regulations specifically note that permits are not required simply to scare away migratory birds other than threatened or endangered species or eagles.

Although the act prohibits take of migratory birds and the FWS issues permits for activities that include take of migratory birds, neither the act nor the regulations address incidental take. Most land development activities disturb some habitat used by one or more species of migratory bird; even the disruption of a lawn could disturb habitat used by American robins, and removal of a parking lot could disturb habitat used by ring-billed gulls. Requiring take permits for routine disturbance of migratory bird habitat would be administratively impractical. In practice, few development activities, government or private sector, receive permits under the act. The act clearly was intended only to regulate activities that directly impact migratory birds. One exception specifically authorizes the "Armed Forces" to "take migratory birds incidental to military readiness activities" as long as the military implements conservation measures to minimize and mitigate adverse effects.[25] This exception constitutes a compromise between the military and some of our nation's most aggressive environmental activists. It represented a compromise between environmental activists and the Department of Defense.

In addition, the FWS has in recent years used the act as the basis for requiring cleanups of contaminated sites potentially affecting migratory birds. In a letter issued in 2011 for a controversial project (which cannot be cited), the FWS stated that the act "prohibits the taking of any migratory birds, their parts, nests, or eggs except as permitted by regulations, and does not require intent to be proven." It remains to be seen whether the FWS's authority to require permits under the act for incidental take of migratory birds will be successfully challenged in court.

4.7 Others

The regulatory requirements discussed next do not as tightly overlap with the Endangered Species Act as those previously discussed, but they are nevertheless substantially interrelated to species and habitats. Space does not allow for separate discussions of every environmental regulatory requirement potentially interrelated with the Endangered Species Act. As stated at the beginning of this chapter, Table 4.1 is not a comprehensive list of every interconnected environmental regulation. Environmental planners need not, however, be bewildered; as repeatedly emphasized in this chapter, early and frequent communication with the Services is the best way to learn of the diverse requirements and integrate their compliance into an efficient and seamless effort.

4.7.1 Executive Order 11988 (Floodplain Management)

It might seem surprising that there are no direct federal permit requirements for development impacts affecting floodplains, those areas usually adjacent to rivers and other surface water bodies that are subject to inundation by floodwaters. The definition of "waters of the United States" subject to Clean Water Act requirements, which includes wetlands, does not include floodplains other than floodplain areas that are also wetlands or surface waters. Of course, minimizing floodplain encroachment is a key objective of environmental planning, not only because flooding can adversely affect developed property, but also because floodwaters deflected by new fill can be displaced to other property that would not have previously been flooded. Floodplains contain many ecologically sensitive lands that provide specialized habitat for threatened and endangered species. Even portions of floodplains that are not wetlands can contain riparian vegetation or coastal vegetation that plays a key role in protecting water quality, providing wildlife habitat, and supporting aquatic and terrestrial food chains.

Although there is no federal floodplain permitting requirement, many states and localities regulate development in floodplains. For example, Maryland requires developers working in floodplains to obtain Waterway Construction Permits from the state. Prior to 1990, the state's efforts to protect nontidal (inland) wetlands were closely linked to the Waterway Construction Act, which was primarily directed at floodplains. Subsequent to enactment of the state's Non-tidal Wetlands Act in 1990, the state has had parallel permitting requirements directed at inland floodplains and wetlands. As another example, New Jersey requires anyone impacting a floodplain to obtain a stream encroachment permit from the state. New Jersey considers the floodplain to be an integral part of the stream itself. A lot of states, especially in the south and west, have no state-level floodplain permitting programs, but many local jurisdictions in those states consider floodplain impacts when deciding on applications for local building permits.

Although there is no requirement for a federal permit to impact floodplains, President Carter did issue Executive Order 11988, Floodplain Management, which directs federal agencies to

> take action to reduce the risk of flood loss, to minimize the impact of floods on human safety, health and welfare, and to restore and preserve the natural and beneficial values served by floodplains in carrying out its responsibilities for (1) acquiring, managing, and disposing of Federal lands, and facilities; (2) providing Federally undertaken, financed, or assisted construction and improvements; and (3) conducting Federal activities and programs affecting land use, including but not limited to water and related land resources planning, regulating, and licensing activities.[26]

The Federal Emergency Management Agency (FEMA), although not empowered to require permits for floodplain development, maps floodplains over many areas of the United States. The maps, termed flood insurance rate maps, primarily serve as the basis for qualifying owners of property for federally supported flood insurance. But, they also serve as useful reference maps, somewhat analogous to the National Wetland Inventory maps for wetlands.

Unlike wetlands, which are defined on the basis of hydrology, soils, and vegetation, floodplains are defined based exclusively on hydrology. Executive Order 11988 defines floodplains as

> the lowland and relatively flat areas adjoining inland and coastal waters including floodprone areas of offshore islands, including at a minimum, that area subject to a one percent or greater chance of flooding in any given year.[27]

This definition is sometimes termed the "100-year floodplain" or "base flood." It is based on the statistical concept of expected value; based on a probability of 0.01 (1%) for flooding in any given year, one may expect one flooding event per 100-year period. The term and definition easily lead to misconception. One flood per 100 years seems to suggest a highly infrequent event. But even if the probability can be substantiated so mathematically, the one flood per 100 years must be interpreted as a long-term average. Two, three, or more flood events might occur over a considerably shorter time interval, perhaps followed by intervals of more than 100 years with no flood. Moreover, our records of historical flood occurrences date back only a few decades, a couple of centuries even in the oldest settled areas. Our empirical basis for developing the 0.01 probability datum is very sketchy; the mathematical elegance obscures the inherent uncertainty in the estimate. And, landscapes change over time. Vegetation and other land cover as well as surface soil properties affected by land use changes play a key role in how precipitation events flow over the land surface and lead to flood events. Forest clearing and urbanization are but two of the most obvious factors influencing how flood probabilities change over time.

Some other terms are commonly encountered in discussions of floodplains. One is the 500-year floodplain, defined in an analogous mathematical format as inclusive of areas expected to flood once per 500 years (probability of flooding greater than 0.02% in a given year). Logically, the 500-year floodplain extends farther landward of the corresponding water body (e.g., river) than the 100-year floodplain. Similarly defined floodplains may be defined for any expected flood interval (or any probability of flooding in a given year). Executive Order 11988 does not address floodplains more expansive than the 100-year floodplain, but many agencies (e.g., the U.S. EPA and Department of Energy) have issued agency-specific regulations that limit certain highly sensitive actions by the agency (e.g., development of hazardous waste facilities) in the 500-year floodplain, sometimes termed the "critical action" floodplain.

Another term used on FIRMs is the floodway. FEMA defined the "regulatory floodway" as the "channel of a river or other watercourse and the adjacent land areas that must be reserved in order to discharge the base flood without cumulatively increasing the water surface elevation more than a designated height."[28] Usually, floodways are defined for the 100-year flood and a height of 1 foot, that is, as the portion of the floodplain that if filled cumulatively from the landward edge toward the channel would raise the 100-year flood elevation by 1 foot. Although defined primarily for flood insurance purposes, the floodway clearly is that part of a floodplain where if disturbed the hydrological impacts would be greatest. Although the sensitivity of riparian habitats and other sensitive habitats frequented by threatened or endangered species generally increases closer to river channels and other shorelines, this is not always the case. The term *floodway* therefore has little meaning in the context of the Endangered Species Act.

The name of the executive order, Floodplain Management, says much about its objectives. Unlike the commonly associated executive order for wetlands, termed Protection of Wetlands (see next paragraph), the name of the floodplain executive order is not Protection of Floodplains. It emphasizes management, not protection per se, of floodplains; it does not seek complete avoidance of floodplain encroachment by new development. It does not establish a permitting or consultation requirement. It merely directs agencies to consider prudent management of floodplains when planning actions. Efforts to minimize floodplain impacts are commonly documented in EISs and other NEPA documents. Because of the specialized habitats commonly present in floodplains for threatened and endangered species, floodplain impacts can also play a role in biological assessments, incidental take permits, habitat conservation plans, and other activities under the Endangered Species Act.

4.7.2 Executive Order 11990 (Protection of Wetlands)

Unlike the better-known and more controversial Clean Water Act Section 404, Executive Order 11990 specifically instructs federal agencies, and not other parties, to minimize impacts to wetlands. Executive Order 11990 does not rely on authority to protect "waters of the United States" but directly addresses wetlands, defined (with minor wording differences) in a manner similar to how wetlands are defined under the Clean Water Act. The order, issued in 1977, does not pre-date Section 404 but does pre-date the widespread consensus that the "waters of the United States" regulated under Section 404 include wetlands. Not only was the order not superfluous at that time, but it was also groundbreaking. It was the first government-wide directive to conserve and protect wetlands, issued at a time when the very concept of "wetlands" was unfamiliar to many agencies and when many agency personnel still regarded wetlands as wastelands that should be eliminated rather than protected. The importance of the order has since been partially eclipsed by Section 404, but key differences, discussed in the following, remain.

The order defines wetlands as

> those areas that are inundated by surface or ground water with a frequency sufficient to support and under normal circumstances does or would support a prevalence of vegetative or aquatic life that requires saturated or seasonally saturated soil conditions for growth and reproduction. Wetlands generally include swamps, marshes, bogs, and similar areas such as sloughs, potholes, wet meadows, river overflows, mud flats, and natural ponds.[29]

In general, areas meeting the definition of wetlands under Section 404 also meet the definition for the order. Separate wetland delineations are not necessary. The order specifically addresses wetlands; most other waters of the United States are not covered. However, the order's definition of wetlands specifically includes some water features such as sloughs, river overflows, and natural ponds that are sometimes referred to as waters of the United States rather than wetlands in the context of Section 404.

Executive Order 11990 establishes requirements only for federal agencies. It does not require a permit for wetland impacts, but merely directs agencies to incorporate wetland protection objectives into their programs and proposals. Specifically, it instructs agencies to

> provide leadership and shall take action to minimize the destruction, loss or degradation of wetlands, and to preserve and enhance the natural and beneficial values of wetlands in carrying out the agency's responsibilities for (1) acquiring, managing, and disposing of Federal lands and facilities; and (2) providing Federally undertaken, financed, or assisted construction and improvements; and (3) conducting Federal activities and programs affecting land use, including but not limited to water and related land resources planning, regulating, and licensing activities.[30]

It generally does not impose limitations on private property use or on the actions of private parties, even indirectly. It clearly states that its requirements do not apply to "the issuance by federal agencies of permits, licenses, or allocations to private parties for activities involving wetlands on non-Federal property." It could, however, possibly limit the ability of a federal agency to acquire nonfederal property containing wetlands for development as part of a federal project. The order is much less prescriptive than Section 404 and much less focused on wetland acreage and "no net loss." The focus is directed much more on wetland values and functions. Section 5 of the order states:

> In carrying out the activities described in Section I of this Order, each agency shall consider factors relevant to a proposal's effect on the survival and quality of the wetlands. Among these factors are:
>
> (a) public health, safety, and welfare, including water supply, quality, recharge and discharge; pollution; flood and storm hazards; and sediment and erosion;
> (b) maintenance of natural systems, including conservation and long term productivity of existing flora and fauna, species and habitat diversity and stability, hydrologic utility, fish, wildlife, timber, and food and fiber resources; and
> (c) other uses of wetlands in the public interest, including recreational, scientific, and cultural uses.[31]

Note the reference to many of the commonly recognized wetland values and functions, including protection and recharge of water sources, flooding and storm hazards, sediment and erosion protection, and species and habitat diversity.

One key difference between the order and Section 404 is that the scope of the order is not limited to the discharge of dredged or fill material. It applies more broadly to any action of federal agencies that adversely affects wetlands, including actions that diminish wetland values and functions even if they do not reduce wetland acreage. The order does not establish a permitting or consultation requirement, although some agencies (e.g., the U.S. Department of Energy) have established internal requirements to document compliance with the order. EISs and EAs for actions affecting wetlands commonly include a description of compliance with the order. Even if no specific documentation is required, agencies must still be cognizant of the order when planning actions.

4.7.3 Executive Order 13112

It is not a perfectly obverse rule to the Endangered Species Act, but if the Endangered Species Act seeks to protect and nurture species that are declining or have otherwise become rare, then Executive Order 13112 seeks to control and limit the spread of species that, instead of having become rare, are proliferating so fast and vigorously as to be undesirable from a conservation perspective. The order direct federal agencies to identify their actions that might encourage the spread of invasive species, to seek to the extent practicable to minimize the spread of invasive species, and to

> not authorize, fund, or carry out actions that it believes are likely to cause or promote the introduction or spread of invasive species in the United States or elsewhere unless, pursuant to guidelines that it has prescribed, the agency has determined and made public its determination that the benefits of such actions clearly outweigh the potential harm caused by invasive species; and that all feasible and prudent measures to minimize risk of harm will be taken in conjunction with the actions.[32]

The trend parallels NEPA and many other environmental regulations and Executive Orders; it does not actually prohibit agencies taking actions that result in the spread of invasive species but instead forces agencies to evaluate how contemplated actions might exacerbate the problems caused by invasive species and consider possible ways to minimize those effects. Since issuance of the order in 1999, many EISs and EAs for actions disturbing natural habitats specifically address possible effects from invasive species. Invasive species may also compete with and adversely affect threatened and endangered species and infest and degrade critical habitats and are hence addressed in many biological assessments and other documents prepared for the Endangered Species Act.

When one considers invasive species, weeds come to mind. Although definitions vary, weeds are generally perceived as plants growing in settings where they are undesirable and may include undesirable plants growing in agricultural settings such as fields or pastures, landscape settings such as lawns and gardens, and industrial settings such as highways and rights-of-way. But, Executive Order 13112 addresses more than just weeds. It defines invasive species as "alien species whose introduction does or is likely to cause economic or environmental harm or harm to human health."[33] Note use of the term *alien*; the order defines an alien species as "with respect to a particular ecosystem, any species, including its seeds, eggs, spores, or other biological material capable of propagating that species, that is not native to that ecosystem."[34] Many agricultural and landscape weeds are of Eurasian origin and hence alien; that combined with their economically detrimental effects on agriculture or suburbia qualify them as invasive under the order. Thus, Japanese honeysuckle (*Lonicera japonica*), an aggressive weedy vine introduced to North America from Japan that can degrade agricultural pastures and young tree stands, qualifies under the definition, but common cattail (*Typha latifolia*), a plant native to most of North America that can choke out more desirable plants from ditches and wetlands, does not qualify. Invasive plant species can detrimentally affect natural settings other than agricultural, suburban, or industrial lands. For example, purple loosestrife (*Lythrum salicaria*), a Eurasian plant species that infests wetlands in many Northeastern states but rarely affects agricultural fields or gardens, is a common target of actions under the order.

Of course, the order and its definition of invasive species extend beyond plants. Some of the worst invasive species are insects. For example, the emerald ash borer (*Agrilus planipennis*) is an exotic beetle that was first observed near Detroit, Michigan, in 2002 and has since killed millions of ash (*Fraxinus*) trees in Michigan, Illinois, Indiana, Kentucky, Minnesota, Missouri, New York, Ohio, Ontario, Pennsylvania, Tennessee, Quebec, Virginia, West Virginia, and Wisconsin. Quarantines have since been established in Michigan, Illinois, Indiana, Iowa, Maryland, Minnesota, Missouri, Ohio, New York, Ontario, Pennsylvania, Tennessee, Virginia, West Virginia, Wisconsin, and Kentucky in a desperate attempt to protect ash trees.[35] The rapid spread and widespread loss of ash trees expanding progressively outward from Detroit since its discovery in 2002 is reminiscent of the rapid and widespread loss of the American chestnut (*Castanea dentata*) since the discovery of the chestnut blight fungal disease near New York City in 1905.

The order does not establish specific permitting or consultation requirements. Agencies specifically handling invasive species have to comply, however, with requirements established by the U.S. Department of Agriculture and some states. EISs and EAs for projects disturbing natural habitats should describe compliance with the order. Information on invasive species is often necessary in the biological resources sections of EISs and EAs, as well as in biological assessments prepared for Section 7, and protection against invasive species is a key element whenever designing natural resource mitigation plans, whether in the context of NEPA, Section 404, or Section 7.

4.7.4 National Historic Preservation Act

The National Historic Preservation Act, enacted in 1966 to protect our nation's historical and archaeological heritage (commonly grouped under the term *cultural resources*), targets a very different set of resources than the Endangered Species Act. It really is related to the Endangered Species Act less than many other environmental regulations not discussed in this chapter. But, Section 106 of the National Historic Preservation Act establishes a consultation process for federal actions that somewhat parallels that under Section 7 of the Endangered Species Act. For this reason, many environmental planners casually refer to Section 106 and Section 7 as the two key interagency environmental consultation requirements for federal actions. Actually, they are not the only consultation requirements (e.g., note the requirement for consultation under the Fish and Wildlife Coordination Act for projects affecting surface water resources), but they are the best known.

The most notable commonalities between the Section 7 and Section 106 consultation processes include

- applicability to actions proposed, funded, permitted, or authorized by federal agencies,
- the need for early and frequent communication between action agencies and specialized resource protection agencies,
- a general division of the effort into informal and formal stages,
- applicability to federal but not nonfederal actions but frequent "federalization" that extends consultation requirements to nonfederal actions requiring federal funding, permits, or approvals, and
- a need for phased investigations of increasing complexity for project sites.

As is true for Section 7, federalization of Section 106 requirements onto private land development projects most often occurs in connection with wetland impacts requiring permits from the U.S. Army Corps of Engineers under Section 404 of the Clean Water Act. Not only do developers commonly have to hire biologists to survey for threatened and endangered species on their proposed project sites, but also they commonly have to hire archaeologists and historians to poke around their sites. There is an intuitive link between threatened and endangered species and wetlands, but any link between wetland impacts and impacts to historical and archaeological resources is less intuitively apparent. But, wetlands are the regulatory "handle" that ties the federal National Historic Preservation Act to many private development projects.

There is one key difference in the Section 106 and Section 7 processes: Consultation under Section 106 is directed to state experts rather than experts at federal "service" agencies. Each state appoints a state historic preservation officer (SHPO; sometimes casually referred to as the "shippo") to carry out the act's objectives in that state, including responding to Section 106 consultation requests. The SHPOs are not federal employees. They work for state agencies. American Indian tribes are authorized to designate tribal historic preservation officers (TPHOs) to function like SHPOs for lands under tribal jurisdiction. As expected, many tribes are keenly interested in protecting cultural resources related to their past and therefore take their responsibilities in the Section 106 process very seriously. Some national guidance is provided by the Federal Advisory Committee on Historic Preservation, but most decisions regarding individual proposed projects are made by the SHPOs and THPOS.

Like Section 7 of the Endangered Species Act, Section 106 is normally integrated into the NEPA process for projects requiring an EIS or EA. However, as for Section 7, exemption from NEPA does not necessarily imply exemption from Section 106. Other than its parallel framework to Section 7, and its status as an environmental regulation, Section 106 has little to do with threatened or endangered species, habitats, or even ecological resources in general. It is therefore not discussed further in this book.

Notes

1. 42 U.S.C. 4321 et seq., the National Environmental Policy Act of 1969, as amended.
2. 42 U.S.C 4322(C).
3. 40 C.F.R. 1500-1508.
4. 40 C.F.R. 1502.25(a).
5. *United States v. Riverside Bayview Homes, Inc.*, 474 U.S. 121 (1985).
6. 33 C.F.R. 328.3(a)(3).
7. 33. C.F.R. 328.3(e).
8. 33 C.F.R. 328.3(b).
9. Environmental Laboratory. 1987. *Corps of Engineers Wetlands Delineation Manual*, Technical Report Y-87-l, U.S. Army Engineer Waterways Experiment Station, Vicksburg, MS.
10. Adamus, P.R., et al. 1987. *Wetland Evaluation Technique (WET), Volume II: Methodology*, National Technical Information Service No. ADA 189968, U.S. Army Corps of Engineers, Waterways Experiment Station, Vicksburg, MS.
11. U.S. Army Corps of Engineers. 1995. *The Highway Methodology Workbook: Supplement to Wetland Functions and Values: A Descriptive Approach*, U.S. Army Corps of Engineers, New England Division, Concord, MA.
12. Clairain, E.J., Jr. 2002. *Hydrogeomorphic Approach to Assessing Wetland Functions: Guidelines for Developing Regional Guidebooks*, ERDC/EL TR-02-3, U.S. Army Corps of Engineers, Engineer Research and Development Center, Vicksburg, MS.
13. Florida Administrative Code, Chapter 62-345, Uniform Mitigation Assessment Method.
14. Bartoldus, C.C. 1999. *A Comprehensive Review of Wetland Assessment Procedures: A Guide for Wetland Practitioners*, Environmental Concern, St. Michaels, MD.
15. 16 U.S.C 661, Fish and Wildlife Coordination Act, Declaration of Purpose; Cooperation of Agencies; Surveys and Investigations; Donations.

16. 16 U.S.C. 668(a).
17. The status of the bald eagle, like other listed species, depends on the state. At various times prior to 1995, the bald eagle was listed as endangered in some states, threatened in others, and not listed in Alaska (where it has remained common) and Hawaii (where the species has never occurred).
18. *Federal Register*, Vol. 72, No. 130, July 9, 2007, 37346–37372.
19. *Federal Register*, Vol. 72, No. 107, June 5, 2007, 31132–31140.
20. U.S. Fish and Wildlife Service. May 2007. *National Bald Eagle Management Guidelines*. 23 pp.
21. U.S. Fish and Wildlife Service, *National Bald Eagle Management Guidelines*, p. 15. Washington, DC: Author.
22. 50 C.F.R. 22.26, permits for eagle take that is associated with, but not the purpose of, an activity.
23. U.S. Fish and Wildlife Service. 2011. Draft Environmental Assessment to Permit Take as Provided Under the Bald and Golden Eagle Protection Act for the West Butte Wind Project, Oregon. Prepared by the Pacific Region, Divisions of Ecological Services and Migratory Birds and State Programs, December 29. Washington, DC: Author.
24. 50 C.F.R. 21 Subpart D.
25. 50 C.F.R. 21.15.
26. Executive Order 11988, Floodplain Management, Section 1.
27. Ibid., Section 6.
28. 44 C.F.R. 59.1, Title 44—Emergency Management and Assistance. Chapter I—Federal Emergency Management Agency. Part 59—General Provisions.
29. Executive Order 11990, Protection of Wetlands, Section 6(c).
30. Ibid., Section 1.
31. Ibid., Section 5.
32. Executive Order 13112, Invasive Species, Section 2(a)(3).
33. Ibid., Section 1.
34. Ibid.
35. U.S. Forest Service. Emerald Ash Borer home page. Available at http://www.emeraldashborer.info. Accessed October 20, 2011.

5

Section 7: The Federal Consultation Process

5.1 Introduction

Section 7 is the part of the Endangered Species Act most commonly encountered by environmental consultants and agency staff who are not biological specialists. It is that part of the Endangered Species Act most intimately associated with the National Environmental Policy Act (NEPA; see Section 4.2) and the overall environmental planning process for federal projects and projects requiring federal permits. The primary goal of the Endangered Species Act is recovery of species recognized as in danger of extinction; careful environmental planning that circumvents the need for incidental take permits promotes recovery far more effectively than issuing such permits. Timely and cautious planning not only can avoid unnecessary adverse effects on threatened and endangered species and critical habitats but also can head off unnecessary delays and costs required to respond to unforeseen consequences of actions. It can also head off the need for mitigation, which can often be considerably more expensive and time consuming than avoiding the impacts in the first place.

For federal projects, Section 7 is the primary venue for environmental planning in the context of the Endangered Species Act. Like NEPA, Section 7 is documented through paperwork, but the success of Section 7 relies not on the quality of the paperwork but on the quality of decisions made in the process of producing the paperwork.[1] The most important elements driving the success of those decisions include

- Availability of relevant data;
- Frequent and effective communication with experts; and
- Careful evaluation of possible impacts and how those impacts may be mitigated.

As for NEPA, Section 7 is a federal process required only for actions of federal agencies, although the involvement of federal agencies in funding, permitting, or approving development projects proposed by private-sector parties frequently entangles those parties in the Section 7 process. The possible applicability of Section 7 to proposed projects is discussed in Section 5.2. As more and better data become available on the species and habitats protected by the act, environmental planners can make decisions that are more informed about proposed actions that might affect those resources. The most comprehensive repositories of technical data on listed species and habitats are maintained by the two federal agencies assigned to administer the Endangered Species Act: the U.S. Fish and Wildlife Service and the National Marine Fisheries Service (FWS and NMFS, collectively referred to as the Services). The availability of data from the Services and other sources is discussed in Section 5.3. The Services also house the most technically specialized expertise within the federal government regarding listed species and habitats. The availability and need for frequent and effective informal communication with the Services is discussed in Section 5.4. The possible need for targeted site-specific field surveys to determine whether listed species or habitats are present in areas potentially subject to effects from the project is discussed in Section 5.5. How federal agencies planning projects can effectively evaluate possible effects on listed resources and report those effects in a biological assessment is discussed in Section 5.6. Although possible conflicts between projects and listed resources can often be avoided once appropriate data and expertise are summoned, some projects cannot be practicably modified to avoid those impacts. For those projects, the goals of the act are achieved through the formal consultation process, discussed in Chapter 6.

5.2 Who Must Comply

The Endangered Species Act establishes that

> Each federal agency shall, in consultation with and with the assistance of the Secretary, insure that any action authorized, funded, or carried out by such agency (hereinafter in this section referred to as an "agency action") is not likely to jeopardize the continued existence of any endangered species or threatened species or result in the destruction or adverse modification of habitat of such species which is determined by the Secretary, after consultation as appropriate with affected States, to be critical, unless such agency has been granted an exemption for such action by the Committee pursuant to subsection (h) of this section. In fulfilling the requirements of this paragraph each agency shall use the best scientific and commercial data available.[2]

The words *carried out* are simple: Actions proposed or sponsored by any federal agency are subject to the interagency consultation requirements of Section 7 if there is any potential for adverse effects to listed resources. The actions do not have to pertain directly to management of biological resources. The effects on listed resources can be "incidental" to the purpose of the action. For example, building an aircraft hangar or runway and carrying out military exercises are not specifically intended to disrupt rare species or habitat. Nevertheless, the Department of Defense agency sponsoring construction of the facilities or carrying out the exercises must comply with Section 7. The fact that protection and management of species and habitats are unrelated to the Department of Defense's mission of protecting the country does not alleviate it of Section 7's requirements.

The word *funded* is also simple. If a federal agency awards or loans funds to a project, the agency must engage in the Section 7 process, regardless of who is primarily responsible for implementing the project. An illustrative example is a recent program of loan guarantees from the U.S. Department of Energy for privately developed energy development projects. The stated purpose of the loan guarantees is to stimulate development of various energy generation facilities that might not be possible if developers had to rely only on traditional sources of financing such as bank loans.[3] Examples of energy generation technologies targeted by the loan guarantee program include wind turbine facilities, solar generation facilities, and nuclear reactors. Most have the potential to affect listed resources adversely, through both habitat losses to develop the facilities and impacts associated with the inherent operations of the facilities once constructed. For example, operation of wind turbine facilities involves moving blades that can strike birds and bats, development of power plants of any type requires large commitments of water and land, and development of electric transmission lines fragments landscapes.

More far reaching is *authorized*, which includes issuance of permits, licenses, and other approvals without which the project may not proceed. Table 5.1 presents examples of federal permits and other authorizations that can require Section 7 consultation. An unexpected conundrum that has emerged repeatedly over the past 40-plus years of implementation of NEPA is the question of how much of a private development project subject to federal permitting or authorization is "federalized" and therefore subject to review under NEPA and other environmental regulatory requirements, including the Section 7 consultation requirements of Endangered Species Act. The question, sometimes termed the "small federal handle problem," has arisen most controversially with respect to the U.S. Army Corps of Engineers (USACE) permitting program for impacts to wetlands and other waters of the United States under Section 404 of the Clean Water Act (described in Section 4.3 of this book). The question usually manifests itself as how much of the project must be assessed by USACE in its NEPA review process before issuing a Section 404 permit—an entire land development project or just those portions of a land development project encroaching into wetlands and other waters of the United States. The opinion of courts has varied considerably, wavering between a philosophy that meaningful NEPA

TABLE 5.1

Examples of Federal Permits and Authorizations Potentially Requiring
Section 7 Consultation

Permit, Authorization, or Other Federalizing Action	Administering Federal Agency
Section 404 permit	U.S. Army Corps of Engineers
Section 10 permit	U.S. Army Corps of Engineers
Federal land easement	Any agency administering land
Federal water allocation	Any agency authorized to allocate water resources
Federal loans or grants	Any agency
Federal loan guarantee	U.S. Department of Energy
FERC license	Federal Energy Regulatory Commission
NRC license	Nuclear Regulatory Commission
CERCLA investigations and cleanups	U.S. Environmental Protection Agency
FCC license	Federal Communications Commission

compliance demands consideration of a complete, integrated project versus a philosophy that permitting agencies are not required (or even authorized) to consider impacts from private development projects that do not fall within the agency's permitting authority.[4]

Section 7 is an interagency consultation process involving one or more agencies proposing to conduct, fund, or authorize an action (the action agency or action agencies) and one or both of the Services. Although some of the larger federal agencies whose decisions most intimately affect land and water resources hire ecologists and other natural resource biologists to work on their staffs (or hire contractors to provide that expertise), the primary source of expertise related to management of listed species is housed within the Services. Although the power to make decisions rests with the action agencies, those agencies are expected to avail themselves of the expertise housed within the Services prior to making decisions. The Services are empowered only to offer advice, not to force action agencies to decide in a certain way. Action agencies are required to seek the Services' advice but not necessarily to make decisions consistent with recommendations in that advice. Action agencies are required to consult with the Services, not seek approval from the Services. Nevertheless, action agencies are expected to make use of the advice and factor it into their decision making. What constitutes adequate consideration of the advice is a matter of interpretation. Action agencies can be sued if outside interests feel as if the Services' advice was not considered prior to making decisions. Action agencies must be prepared to justify the rationale for decisions that appear to be in contradiction to recommendations proffered by the Services through the consultation process.

The motivation deriving from the threat of lawsuits is a common theme in environmental planning in the United States. It is most evident in NEPA (see Section 4.2), where each agency is empowered to self-enforce its own compliance. The Council on Environmental Quality (CEQ) offers guidance on NEPA but lacks enforcement authority. The Environmental Protection Agency (EPA) has likewise developed specific general guidance documents on NEPA that compliment—not contradict—CEQ guidance and receive copies of each published environmental impact statements (EISs). But, EPA also lacks direct NEPA enforcement power. An interesting balance of power exists to drive enforcement of Section 404 of the Clean Water Act. The agency empowered to administer the Section 404 permitting program of Section 404 directly is the USACE. However, the agency primarily empowered to administer the rest of the Clean Water Act is the EPA. The USACE had been administering the older Rivers and Harbors Act permitting program since the nineteenth century, and extending the older program to cover the newer interrelated permit requirements under Section 404 made sense. To involve EPA, the framers of the Clean Water Act granted EPA "veto authority" to block USACE from issuing Section 404 permits determined not to meet the spirit and intent of the Clean Water Act.[5] Even though EPA rarely uses its Section 404 veto authority,[6] the possibility of a veto helps to motivate USACE only to issue permits not likely to trigger a veto. EPA does not need a lawsuit in order to impose its Section 404 veto authority.

5.3 Basic Information Sources

In the early years of the Section 7 process, action agency staff or consultants often initiated their effort by communicating with the Services through the informal processes described in Section 5.4. As stressed in Section 5.4, communication with the Services as early as possible in the project planning process is extremely important to focus further consideration of listed species and habitats properly and avoid unnecessary evaluation efforts. But, action agencies now have easy access to several sources of information on threatened and endangered species that were formerly unavailable or inaccessible. Agency representatives can readily educate themselves on the possible occurrence of listed resources in the general region of their anticipated projects even prior to contacting the Services. Researching these easily accessible data sources is not a substitute for the informal consultation process described in Section 5.4 or for early or frequent communication with the Services. But, reviewing these sources as early as possible can better position agency representatives to more effectively communicate with the Services and understand the information that the Services provide. Moreover, early and immediate access to these data sources can help agencies avoid obvious and easily avoided conflicts with threatened and endangered species when initially formulating alternatives.

5.3.1 Web Sites

In recent years, a large amount of data on threatened and endangered species and other listed resources has been posted on Web sites. Web sites are, however, constantly changing, and new and reconfigured Web sites are frequently introduced. Use of the information from Web sites is not a substitute for individualized research. Many Web sites are intelligible to project managers and environmental consultants who are not biologists or possess degrees in biology or related disciplines. However, generalists working for action agencies should not hesitate to seek assistance from knowledgeable biologists as necessary to understand information contained in Web sites.

The national Web site for the FWS Endangered Species Program provides a searchable database of listed threatened and endangered species and critical habitats potentially occurring in individual states and counties.[7] The marine aquatic species and habitats covered by the NMFS are generally by their nature less localized. The NMFS national Web site does not presently include state or county lists, although the Web sites of some of the NMFS regions provide more regionalized information. For example, the Southeast Regional Office presents state lists.[8] The limitations of data provided for lists for entire states and counties must be understood. States and counties are political entities whose boundaries are not defined based on biological or ecological parameters. All states and nearly all counties comprise landscapes consisting of multiple habitat types occurring in multiple physiographic settings. Furthermore, the possible occurrence of a species in a state or county does not necessarily indicate that it may occur in a specific portion of that state or county.

Generalists must be extremely careful whenever assuming that a species included in a state or county list is unlikely to occur on or near a given site or route. However, generalists can sometimes deduce that a species requiring a specialized habitat type occurring only in a distinct portion of the state or county is highly unlikely to be affected by certain sites or routes. Experienced biologists may be able to further winnow the number of species and other listed resources subject to possible impact from a given site or route. Project managers who are not biologists may be able to substantially focus their efforts by conferring with biologists, either on staff or with consulting firms.

Section 5.4 also discusses the Natural Heritage Programs administered by individual states. Until recently, action agencies and their consultants could access Natural Heritage Program data in most states only by writing. Some states still require written requests, and some charge for a response. Some states have, however, published substantial Natural Heritage data on state Web sites. For example, the program for Massachusetts publishes lists of federal and state-listed species for each town[9] (towns in Massachusetts are subdivisions of counties).

Information from Web sites and other readily available data sources is especially important when initially conceptualizing a project. Large numbers of possible locations for a project can be rapidly screened by action agency generalists without having to seek assistance from specialized experts and without having to wait for input from the Services. Examples of locations might be sites for facilities such as power plants, mines, or industrial parks or routes for linear facilities such as highways, pipelines, and electrical transmission lines. If the databases reveal that a possible site or route is located at or close to a known occurrence of a listed species or habitat, the action agency may consider eliminating the location from further consideration to avoid challenges of Endangered Species Act compliance. However, the siting process for most projects must consider factors other than those encompassed by the Endangered Species Act, including nonenvironmental factors such as property availability, development costs, and proximity to existing transportation routes, population centers, and support facilities.

5.3.2 Other Data Sources

Action agencies can access some data maintained by the Services even before they initiate the more regimented processes described in Section 5.4 as "informal consultation." The Services encourage project planners to communicate early and often regarding the potential occurrence of protected resources. Written requests are usually necessary, although representatives in some field offices may be willing to discuss projects informally over the telephone. The staff at most service offices are, however, usually very busy. It is therefore sometimes helpful to follow a written request by a telephone call or an e-mail to ensure that a written request does not languish in an inbox.

Many national parks, national forests, military bases, and other federal properties have been the target of past biological survey efforts that have contributed records of listed species and habitats on those properties. The same is true of many established power plant and energy development sites, existing highway and utility rights-of-way, and other properties in state or private ownership that have had to conduct biological surveys to comply with federal or state permitting processes. Many federal and institutional properties have developed geographic information system (GIS) layers (shapefiles) displaying known occurrences of listed resources. However, many properties, especially many privately owned undeveloped tracts of land without a history of environmental survey or permitting, have never been the subject of biological surveys. The fact that the Services do not have any records of known sightings of listed species or habitats on those properties is therefore of little informative value. The Services have long emphasized in their responses to information requests that the absence of recorded occurrences of listed resources in a given area does not necessarily indicate that such resources are not actually present. The Services have however developed lists of potentially occurring listed species and habitats for individual counties (and other geographic localities) and will usually encourage project planners to consider the possible occurrence of each species or habitat type potentially occurring in the affected counties.

As stated in the previous paragraph, most federal properties encompassing substantial areas of natural habitat have at least some history of past survey for biological resources. A quick review of past survey data for these properties is sometimes very helpful before contacting the Services. For example, many military bases have developed integrated natural resource management plans providing comprehensive baseline data on biological resources within their boundaries, including but not limited to threatened and endangered species and critical habitats. Most national forests have similar plans, as do the U.S. Department of Energy national laboratory properties. Frequently, those plans are tied to installation-wide EISs that provide NEPA documentation for ongoing site activities. Past project-specific EISs and other NEPA documents, especially some longer environmental assessments (EAs), may also provide data on listed resources for portions of an installation. Users of past NEPA documents must take careful note of any limitations with respect to the scope of the baseline data provided; data for one past project site cannot be extrapolated as providing comprehensive data for an entire installation.

Users of past data need to consider fully the data's limitations. The geographic scope of the data was discussed in a previous paragraph. Equally important is the age of the data. Biological resources, like other environmental resources, are continually changing. Habitats constantly change through the process of natural succession; fields change into scrub, and scrub changes into forest. Streams and rivers change course, beaches and shorelines accrete and recede, wetlands and springs dry up or become submerged. Development encroaches into once naturally vegetated areas, while abandoned buildings are razed or fall down, and abandoned lands regrow natural vegetation. Even excellent data collected in the past slowly become obsolete. No hard and fast rules are available. Biologists need to use their own scientific knowledge to evaluate the utility of old data.

5.4 Informal Consultation

Informal consultation is just that—unstructured communication between an action agency and the Services regarding the possible effects of a proposal on listed species or critical habitats. But, as a government agency, the FWS has still developed a "formal" definition for "informal" consultation, stating that it is

> an optional process that includes all discussions and correspondence between the Services and a Federal agency or designated non-Federal representative, prior to formal consultation, to determine whether a

proposed Federal action may affect listed species or critical habitat. The process allows the Federal agency to utilize the Services' expertise to evaluate the agency's assessment of potential effects or to suggest possible modifications to the proposed action which could avoid potentially adverse effects.[10]

Although the formal consultation process is better known, most environmental consultants encounter the informal process much more frequently. The Services often inform action agencies during the informal process that formal consultation is not required, using a letter often referred to as a "no further action" letter. Such letters are common for small actions or actions that do not encroach onto significant areas of natural habitat. Receipt of such a letter terminates the Section 7 process. The letters are commonly included as attachments to EISs, EAs, or other environmental planning documents. However, the letters usually advise project proponents that further communication may be required should the project change or new information becomes available. Any changes in design parameters potentially affecting biological resources could trigger the need for further Section 7 communication, such as

- Spatial modifications to the proposed extent of land or water disturbance;
- Increases in noise generation, with respect to intensity (loudness) or frequency;
- Increases in quantities or composition of air emissions or water discharges;
- Changes in the timing or schedule of proposed activities;
- Changes in water withdrawal sources or quantities;
- Increases in structure heights; or
- Additions of new structures or areas of disturbance, even if seemingly small.

From the perspective of action agencies, perhaps the most frustrating limitation to no further action letters is the fact that the Services usually retain the right to require further communication if new information becomes available. If the Services become aware of new evidence that listed species or habitats might be present in the region of a project, they retain the ability to require new communication from the action agency. A no further action letter is therefore a statement that no further effort is required from the action agency—unless someone decides that further effort is required. Although infrequent, this loophole robs action agencies receiving no further action letters to use the letters to claim firm closure on Section 7.

Because the Services issue so many no further action letters, some environmental consultants and project managers routinely handle informal consultations without ever having to participate in a full-blown formal Section

7 consultation. Of course, these persons still need to understand the basic elements of the formal process to participate most effectively in the informal processes. The informal processes exist because of the possible requirement to participate in formal consultation. The Services established the informal processes as an efficiency measure. Use of the informal processes can save both the Services and the action agencies from having to expend resources preparing for a possible but unneccessary formal consultation. The limited staff resources of both are best directed at projects having a true potential for significant adverse effects on listed species and habitats.

Although the FWS describes the informal process as "optional," experienced environmental planners and consultants consider it an essential opportunity to avoid unnecessary formal consultation efforts and efficiently plan those efforts if they are necessary. Action agencies should generally begin informally communicating with the Services as early as possible in the environmental planning process. At a minimum, action agencies should consider known locations for, and potential occurrence of, listed resources whenever considering possible sites for a contemplated action. Likewise, prospective applicants for federal permits should consider known and possible occurrence of listed resources when initially considering sites for their future projects. It is hard to understate how much easier it is to reject a possible site early in the planning process instead of after considerable resources have been spent on it. Many action agency project managers, and especially project engineers in private-sector parties requiring federal permits, do not understand the importance of early consideration of the Endangered Species Act in the overall project planning effort; it is incumbent on their environmental specialists or consultants to do so.

The ready availability of Natural Heritage data, especially as currently available in GIS layers, makes it easy to know whether listed resources are known to occur on possible sites. But, environmental consultants need to avoid the temptation of this simplicity. Presence or absence data is, of course, extremely valuable. Consultants also need to consider possible effects at a distance (e.g., hydrological effects, noise, and effects on migration). The less directly relevant knowledge and experience the consultant possesses, the more important early and frequent communication with the Services is.

The process need not be, and ideally should not be, a one-shot deal. Action agencies are not limited to a single communication prior to formal consultation. Yes, the Services are perennially understaffed. But, they understand how informed early planning can avoid complex problems that drain resources later. Experienced environmental consultants and planners know how to get the attention of the Services even when their staffs are overworked and underfunded, and they develop effective working relationships with Service biologists that carry them through the informal and (if necessary) formal Section 7 consultation processes.

As with other technical procedures, the informal consultation processes are most efficiently performed by knowledgeable biologists, ecologists, or other environmental scientists. But, at least initially, the informal processes may be initiated by project managers or other generalists without specialized biological expertise. As a two-way communication process between the action agency and the Services, the interchange will always involve relatively knowledgeable experts working for the services. Communication can begin with generalists presenting conceptual or preliminary project design information to the Services' experts, who may respond by asking questions about the design and possible environmental effects. Depending on the complexity of the questions, the action agencies may have to engage specialized experts, either on their own staffs or through contracting.

Although the Services frequently issue no further action letters for simple projects following early communication with action agencies, they sometimes need substantial additional information before concluding whether formal consultation is required. They may need targeted surveys of the project area to determine whether listed species or habitats are present (Section 5.5), and they may need an evaluation, termed a biological assessment, of whether and how the project might affect those resources (Section 5.6). Action agencies typically must engage experts to conduct targeted surveys and biological assessments.

5.5 Targeted Surveys

Sometimes field survey efforts are needed, either because all or portions of the affected area have never been previously surveyed or because previous surveys are obsolete, incomplete, or otherwise insufficient. Action agencies can avoid unnecessary effort by carefully considering the need for a survey and, if needed, how to perform a survey most efficiently before initiating survey work (or authorizing a consultant to initiate an effort). Unnecessary effort can also be avoided by never initiating a targeted survey before discussing it with the Services. Effective project management not only avoids wasting money or other project resources on unnecessary or improperly planned surveys but also ensures that any needed surveys are complete in time to complete any required Section 7 consultations and obtain any necessary take permits (usually incidental take permits) without delaying the project schedule. It cannot be emphasized enough that, unlike some environmental survey efforts, biological surveys for many species can only be performed at specific times of the year. These times may coincide with nesting seasons, migration timing, hydrological conditions, or, for plants, flowering seasons. The annual windows of opportunity can be very short for many species, sometimes less than a month, most notably for plants with very brief flowering periods. The windows of opportunity can differ for different species potentially occurring on a given site, thereby requiring multiple survey

TABLE 5.2

Surveys for Federally Listed Plant Species: Proposed Levy Nuclear Power Plant
Units 1 and 2, Levy County, Florida

Species*	Survey Date	Counties	FLUCFCS Codes	Results
Brooksville bellflower (*Campanula robinsiae*) (E)	March 2011	Hernando, Hillsborough	520, 641, 643, 644, 653, 621	No individuals observed
Britton's beargrass (*Nolina brittoniana*) (E)	March 2011	Lake, Hernando, Marion, Polk	412, 413, 421, 427, 432	No individuals observed, but one individual had been observed in common right-of-way in 2009
Florida bonamia (*Bonamia grandiflora*) (T)	July 2011	Lake, Hillsborough, Marion, Polk	412, 413, 421, 432	No individuals observed
Florida golden aster (*Chrysopsis floridana*) (E)	October 2011	Hillsborough, Pinellas	211, 212, 320, 321, 412, 413, 421, 432	No individuals observed
Long-spurred mint (*Dicerandra comutissima*) (E)	October 2011	Marion, Sumter	412, 413, 421, 432	No individuals observed
Cooley's water-willow (*Justicia cooleyi*) (E)	October 2011	Hernando, Sumter	414, 423, 425, 431, 434, 438, 615, 617, 630	No individuals observed

* E = endangered, T = threatened (Federal).

efforts over the course of a year. Table 5.2 illustrates the sequence of survey efforts for threatened and endangered plant species over the course of
a spring, summer, and fall to support a biological assessment and Section 7
consultation for a proposed power plant project in Florida.[11]

The Services have developed suggested field survey methodologies for
several threatened and endangered species commonly addressed in Section
7 and Endangered Species Act permitting efforts. Surveys following these
methodologies are sometimes termed protocol surveys.

The protocols for some species involve a bifurcated effort involving an
initial "Phase 1" effort, the results of which determine whether a more
intensive follow-up "Phase 2" effort is warranted. The Phase 1 effort may
involve simply characterizing the affected habitat, while the Phase 2 effort
may involve attempting actually to observe the targeted species. The timing for the Phase 1 effort may be relatively flexible, although not unlimited,
whereas the Phase 2 effort may have a relatively narrow timing requirement.
While Phase 1 surveys may require only a single visit to only a representative part of the affected habitat, Phase 2 surveys sometimes require multiple

site visits involving observations over all or most of the affected habitats. An example of such a bifurcated survey protocol is the one established by FWS for the bog turtle.[12] The Phase 1 survey involves a single site visit any time of the year to determine whether potentially suitable habitat characteristics are present. Phase 2 involves multiple site visits during a narrowly defined season to inspect potentially suitable habitat areas identified by the Phase 1 process for actual bog turtle individuals.

Protocol surveys are usually conducted by qualified personnel with specialized knowledge about the targeted species. That knowledge involves how to recognize not only the targeted species but also its habitat requirements, behavior, life history, and optimal times and conditions for encountering individuals. The Services maintain lists of qualified surveyors for many species most commonly addressed in Section 7 consultations and other Endangered Species Act compliance activities. Many of the specialists are researchers affiliated with universities or are independent contractors not affiliated with large consulting firms. Generalists assigned to oversee the Endangered Species Act consultation and permitting process or write biological resources sections of EISs or other NEPA documents may not be qualified to perform the requisite protocol surveys for their projects. Even if they possess degrees and other credentials in the biological and ecological sciences, they may have to seek assistance from specialists. Generalists working on consulting contracts may have to subcontract with specialists. No one should embark on protocol surveys without first presenting the proposed survey plans to the Services, including the qualifications of the proposed surveyor(s). Multiple specialists, working on multiple contracts or subcontracts, may be needed for projects affecting more than one listed species.

Surveys for plants typically involve inspection of suitable habitat during times of the year when they are expected to be identifiable. Surveys, performed by a competent observer at an appropriate time of the year, may be

- Comprehensive, involving inspection of all potentially affected habitat;
- Deterministic, involving inspection of those areas of potentially affected habitat determined by the competent observer to be most likely to contain the targeted species; or
- Representative, involving inspection of randomly located sample areas or points within potentially affected habitat.

Comprehensive surveys generally have the least potential for overlooking individuals of the targeted species. They typically require little or no additional effort compared to deterministic or representative surveys for small projects or projects potentially affecting only a small amount of suitable habitat for the targeted species. Limited budgets, tight survey windows, and other practicality issues may prevent comprehensive surveys, forcing observers to inspect only a subset of the potentially affected habitat. Highly experienced observers may be able to identify in advance those portions of the affected habitat with the highest probability for finding the targeted species. Otherwise, it may be necessary to resort to representative sampling. Sampling may be used to identify random representative plots or observation points. Sample plots may be of various shapes and sizes and are sometimes termed quadrats, regardless of shape. The number and spacing of plots or observation points typically rely on statistical considerations. Observations may also be collected by walking or otherwise traversing representative transects (lines or belts) of the investigated area. Several available approaches for sampling vegetation were discussed by Mueller-Dombois and Ellenberg.[13]

Targeted surveys for mobile wildlife can be more complex. Even if suitable habitat is visited by a competent observer at an appropriate time, individuals of the targeted survey may by chance not be present. Even if the observer comprehensively inspects all suitable habitat areas, he or she may still miss targeted individuals that might be present. Field surveys for mobile wildlife may involve spending measured time intervals at representative stations (points) or walking representative transects recording individuals seen or heard. In addition to such direct observations, surveyors commonly look for signs indicative of the occurrence of a species. Examples of wildlife signs include tracks (footprints), abandoned nests or dens, egg debris, and fecal material (scat). Experts are often able to identify the species causing scratches or other disturbances of the soil surface, leaf litter, or vegetation. Surveys can also involve setting baited traps or placing motion-activated cameras. Surveys targeting flying wildlife such as birds and bats sometimes use mist nets, large pieces of nearly transparent netting that capture individuals. Surveys targeting fish species can involve seines, to capture individuals (sort of like an aquatic version of a mist net), or electrofishing, using low electric charges to stun and capture fish. All of these methods allow for releasing the specimens after data are collected.

Survey activities that involve possible physical disturbance to wildlife or fish, such as trapping, mist netting, seining, and electrofishing, usually require scientific collection permits from the FWS and other federal and state agencies. Permits are usually issued only to experienced specialists. Surveys using these methods must be carefully planned to minimize possible injury to individuals of the targeted species as well as individuals of other species that might be incidentally captured.

5.6 Biological Assessments

From the perspective of action agency staff and environmental consultants assigned to oversee compliance with the Endangered Species Act, the biological assessment is to Section 7 as the EIS is to NEPA. The analogy is not perfect. A biological assessment is a document submitted from an action agency to the Services; the document presenting the public face of the Section 7 process is the biological opinion prepared by the Services.

5.6.1 Definition and Overview of the Biological Assessment

The text of the Endangered Species Act and associated regulations do not define a biological assessment. The handbook defines a biological assessment as

> information prepared by, or under the direction of, a Federal agency to determine whether a proposed action is likely to (1) adversely affect listed species or designated critical habitat; (2) jeopardize the continued existence of species that are proposed for listing, or (3) adversely modify proposed critical habitat.[14]

Although action agencies and their consultants commonly prepare biological assessments as neatly packaged reports that have the general appearance of a short EIS or detailed EA focused on listed species and habitats, they are essentially information documents provided to the Services. Although Section 7 is a consultation, not permitting, process, a biological assessment may be considered to be a part of an "application package" to the Services requesting a consultation. A biological assessment is similar in many ways to the "environmental reports" required by some agencies as part of an application for a permit or license; an excellent example is the environmental report that the U.S. Nuclear Regulatory Commission (NRC) requires from applicants for a license to construct or operate a nuclear power plant.[15] The NRC does not use the environmental report as its EIS, but it does use, after independent verification, information contained in the environmental report to prepare its EIS. Similarly, the Services use information from biological assessments to prepare biological opinions. As for the NRC, the Services do not simply parrot the biological assessment in their biological opinions. They must make the information from the biological assessment their own, assuming responsibility for its accuracy.

A biological assessment is a specialized technical document. It is written by technical experts, typically biologists or ecologists, working for action agencies for use by other technical experts working for the Services. Although biological assessments may be made available to the public, they are not primarily public documents. Thus, unlike EISs and EAs, they need not be written in language readily understood by the public. In this respect, they are no different from many of the other technical documents commonly supporting EISs, such as air and water modeling studies. Nobody insists that environmental modeling studies be intelligible to the general public, although the language used in an EIS discussing those studies should be intelligible. In general, the experts working for the Services have more specialized technical knowledge on the subject species than the experts working for the action agencies preparing the biological assessment. The action agencies need not worry about overwhelming the biological expertise possessed by the Services. In fact, the real challenge is to communicate effectively with even more knowledgeable experts.

Nevertheless, even technical documents can be improved if their text is concise, well organized, and effectively presented. Careful writing of biological assessments is important because

- The Services may return a biological assessment that is difficult to read and comprehend without issuing a response;
- The Services may issue requests for additional information or clarification, requiring an interim response from the action agency before issuing a response;
- The Services may make unnecessarily conservative assumptions, imposing additional conditions or mitigation requirements that they might not have if they had a better understanding of the project;
- The Services may take longer to respond if they have to wade through poorly presented or superfluous information; and
- An unnecessarily complicated biological assessment might cause the Services to issue a correspondingly complicated biological opinion.

A biological assessment is a reflection of the proposed action. A concise and well-written biological assessment suggests a carefully planned action; a disorganized biological assessment suggests a poorly planned project. Furthermore, even though a biological assessment does not primarily function as a public presentation of a project, members of the public can usually find access to it. If all else fails, an aggressive member of the public (or activist group) can access a biological assessment, like other government documents, through the Freedom of Information Act.[16]

5.6.2 Project Description in a Biological Assessment

Providing the Services with a well-presented description of the proposed action is perhaps the most important function of a biological assessment. The Services' staffs have thorough biological expertise, usually greater expertise than that on action agency staffs or even their consultants. But, the ability of the Services to understand the project is limited by the information provided by the action agency. Action agencies must describe what they propose to do in enough detail to enable the Services to understand the possible effects on listed resources. Project description information is usually presented in a biological assessment in two ways: as text and as drawings or graphics.

Effective use of graphics can dramatically reduce the length and complexity of text. The old saying that a picture is worth a thousand words applies doubly to environmental reports such as biological assessments. Examples of graphics that can enhance a biological assessment include the following:

- *General Location Figure*: Every report, including biological assessments, should include a figure depicting the general location of the action area. Although the figure need not be precise, it does need to be meaningful. For this broad-level figure, the action area may be depicted as a large star or dot on an outline of a state, county, city, or other widely recognized geographic area. Adding at least a few major landmarks such as larger rivers, highways, or cities will make the map easier to understand. To consolidate the document, this broad-level figure can be presented as an inset within another more detailed figure.

- *Site Boundaries*: A figure depicting the boundaries of the action area is essential. The scale should be detailed enough to depict the action area boundaries in relation to locally relevant geographic features such as streams, rivers, highways, cities, and towns. Most action areas can be depicted as polygons. Action areas for linear projects such as new roads, pipelines, or electric transmission lines are best depicted as long and rectangular, not as lines. Although keeping site boundary figures to a single page is desirable, meaningful figures for large projects or long linear development projects often require multiple pages. Preparers must bear in mind that an action area is not necessarily coterminous with site or right-of-way boundaries or with proposed construction footprint (edge of proposed ground disturbance) boundaries. However, these boundaries should still be depicted internal to the action area boundary.

- *Habitat Map*: A figure depicting the spatial boundaries of habitats in the action area is essential to most biological assessments. Action agencies should confer informally with the Services to receive direction regarding the appropriate content and scale of the habitat map. The level of detail depends on the action. Simple habitat maps using broadly defined habitats such as deciduous forest, evergreen forest, scrub, grassland, and developed areas may be adequate for some simple actions. More detailed habitat definition and naming may be needed for more complex projects.

- *Wetland Map*: Wetlands are a habitat type with special scientific and regulatory meaning. Federal regulation of wetlands is discussed in Section 4.3, and many states regulate wetlands separately from the federal government. Although wetlands are habitats, wetland boundaries are best presented as a separate overlay over other habitat features such as forest, scrub, and grassland. In that way, forested wetlands are distinguished from forested nonwetlands (uplands), grassland wetlands from upland grasslands, and so forth.

The figure should state whether wetland boundaries have been formally delineated and include a reference to the delineation report. If the delineated figures have been formally verified by the USACE or other regulatory agency, the report should also cite the jurisdictional determination or other regulatory verification documentation. If site-specific delineations are not available, approximate wetland mapping is available online through National Wetland Inventory mapping for most areas of the United States.

- *Floodplain Map*: Floodplains, discussed in more detail in Section 4.7, are primarily a hydrological feature, but the presence of a habitat within a floodplain can greatly influence its species composition and potential value to given species. Floodplain mapping is readily available online through the Flood Emergency Management Agency. However, the maps must be properly interpreted (see Section 4.7). Ideally, a biological assessment should explain in its text how the position of a habitat in a floodplain may influence its relation to the subject species.

- *Photographs*: Old-fashioned snapshots can be an invaluable addition to any ecological report, including biological assessments. Of course, today, it is easier than ever to take pictures digitally and incorporate them into text. Photos should be selectively and strategically chosen. Pictures that illustrate site-specific attributes of habitats are usually helpful; close-ups of individual plants and animals often are not. Close-ups of difficult-to-identify or questionable individuals can help in identifying species; however, the use of photographs and other tools in species identification is best performed in early informal stages of Section 7 consultation and not deferred until the biological assessment.

Few actions are completely designed prior to initiation of the Section 7 process, especially at the informal stages. Early communication with the Services can be useful even if only rough, conceptual sketches are available. Even projects undergoing formal Section 7 consultation need not be at a final 100% stage of engineering and design.

5.6.3 Description of Potentially Affected Species and Habitats

Biological assessments need to provide some basic life history information about the species and habitats being evaluated. As discussed in Chapter 2, life history is a description of how individuals of a species grow, feed, behave, reproduce, and senesce. Most species and habitats have been the subject of extensive investigations prior to being listed. Rarely do action agencies or their consultants have to undertake exhaustive research efforts to provide adequate background description. If the Services have produced a recovery plan for a subject species, that plan is usually the best "go-to" source for life history information. Most recovery plans are the product of comprehensive literature reviews and have been thoroughly referenced and peer reviewed. Rarely does a biological assessment have to look beyond an available recovery plan for basic life history information. An exception might be if significant new research findings on the subject species' life history have been published. The value of research beyond the recovery plan is determined by the professional judgment of the preparer.

Recovery plans have not been produced for every listed species. The FWS Web site does, however, have "species profiles" for many listed species, including many species lacking a recovery plan. A species profile is less comprehensive and less scholarly than a recovery plan but is still produced by the FWS and is usually concise and easy to use. Recovery plans are still, however, the best and most authoritative primary source of technical information about a listed species.

The preparers of the biological assessment must use professional judgment regarding the reliability of other information sources not provided by the Services. Peer-reviewed journal articles or other publications are usually preferable, although use of other publications such as books and government publications should not be ruled out. As a last resort, try typing the name of the species into the search line of an Internet search engine. Carefully consider the authenticity of any sources of information found by the search engine. Web sites associated with academic institutions and federal and state natural resources agencies are generally more reliable. Even with these searches, try to look for the primary sources cited or relied on by the Web page. Blogs are nothing more than expressions of opinion and should never be relied on. Online encyclopedia articles on the species of interest can be of questionable reliability if used on their own, but they can be pointers to useful primary sources of information, usually cited at the end of the article.

Life history descriptions in many biological assessments (and biological opinions) are ponderously long and often needlessly so. They should focus on those elements of life history most relevant to an evaluation of how the proposed action might affect the recovery of the target species. Lengthy morphological descriptions are rarely necessary; a biological assessment is not a field identification tool. Application of Charles Eccleston's "sufficiency test"[17] to biological assessments is a good idea. The test was written for application to EISs and other NEPA documents. It is particularly applicable to the "affected environment" or "environmental setting" chapters of an EIS. The basic premise is that each piece of information presented in an EIS should be relevant to the decisions to be made using that EIS. Similarly, each piece of information presented in a biological assessment should support the Services in issuing either a concurrence letter or biological opinion and, if necessary, an incidental take statement (see Chapter 6 for discussion of biological opinions and incidental take statements).

Some examples of pieces of life history information and other general biological background information for a species potentially relevant to a biological assessment include

- Historic range
- Current range
- Suitable habitat
- Preferred habitat
- Estimates of historic and current population levels
- Reasons and possible reasons for population declines
- Listing history (*Federal Register* notices and dates)
- Feeding sources (e.g., herbivore, insectivore, carnivore, omnivore, piscivore, etc.)
- Breeding range
- Wintering (or nonbreeding) range
- Migration seasons
- Migration routes
- Breeding season
- Favored conditions for breeding
- Fecundity (e.g., typical clutch or litter size)
- Normal survival rate of young
- Age of maturity
- Flocking behavior
- Typical life expectancy

The discussion of each of these topics may be brief. Only those topics considered relevant to impact assessment need be discussed. Tables may be a good, concise format for presenting this information.

5.6.4 Impact Assessment in a Biological Assessment

The core of a biological assessment is an evaluation of potential effects (or impacts) on listed species and habitats. Closely paralleling the definition of effects established for NEPA by the CEQ, the handbook recognizes three basic categories of effects: direct, indirect, and cumulative. The concept of direct effects is simple and intuitive. The handbook does not formally define direct effects but describes them as the "immediate effects of the project on the species or its habitat."[18] The CEQ defines direct effects as those "which are caused by the action and occur at the same time and place."[19] Both descriptions identify direct effects as being inherent to the proposed action, connected to the proposed action with respect to both place and time. Impacts to listed species commonly consist of disturbing habitats, converting habitats to developed uses (e.g., a highway, surface mine, ski resort, or reservoir); altering habitat (e.g., cutting timber, thereby converting forest habitat to open habitat); or disturbing habitat (e.g., generating noise or dust). Assessors of biological impacts must individually weigh the possible effects of a proposed action on the biological resources in the affected area. The following considerations may be relevant:

- Rarely is killing or harming a listed species an objective of an action; adverse effects on listed species are usually incidental to other objectives (e.g., building a facility such as a highway or exploitation of resources such as minerals or timber).

- For immobile species such as plants or weakly mobile species such as many small amphibians and reptiles, habitat disturbance generally equates to killing individuals, but even relatively mobile species such as many adult mammals and birds can be adversely affected by loss or degradation of habitat.

- Assessing biological impacts requires an understanding of the life cycle and reproductive habits of the affected species. Adults of many bird and mammal species may be capable of moving out of the way of a disturbance, but eggs and young may not be. Noises that might be little noticed by adults might interfere with breeding or nesting.

- Assessing impacts requires an understanding of food webs and predator and prey relationships. Adverse effects on a species food source can have adverse effects on species relying on that food source.

- Not all possible adverse effects are limited to a spatially defined project site, property political boundary, or footprint of disturbance. Noise, erosion, dust, and runoff and surface flow diversion are examples of activities that can adversely affect species in otherwise undisturbed habitats distant from project activities. Activities on a national forest, military base, or a high right-of-way can affect "off-site" habitats in the surrounding landscape.

- Assessing impacts to mobile species requires consideration of the entire landscape in which they are mobile. The most obvious example involves migratory birds, which can be affected by disturbances in their summer (breeding) habitat, winter (nonbreeding) habitat, and the pathways of migration connecting those habitats. But, most nonmigratory birds and other animals also move about a localized landscape in predictable patterns within an area sometimes referred to as their "home range." Most high-level predators (e.g., hawks and other raptors) establish and defend territories and are affected by disturbances to those territories.

- Assessing impacts to mobile species requires an understanding of carrying capacity. An all-too-common claim in many environmental impact documents is that mobile adults would simply move away from the project site. Such displacement results in those individuals occupying habitat that they would not otherwise have. If the receiving habitat already supports a population exceeding carrying capacity, both the displaced individuals and the home individuals would experience increased competition for the same resources. Some individuals would suffer from fewer resources, ultimately leading to decreased reproduction or increased mortality.

Guidance developed by the CEQ for assessing the significance of impacts in the context of NEPA can be helpful in evaluating potential impacts to listed resources in the context of Section 7. CEQ states that evaluation of the significance of an environmental impact should consider context and intensity, as follows:

> (a) Context. This means that the significance of an action must be analyzed in several contexts such as society as a whole (human, national), the affected region, the affected interests, and the locality. Significance varies with the setting of the proposed action. For instance, in the case of a site-specific action, significance would usually depend upon the effects in the locale rather than in the world as a whole. Both short- and long-term effects are relevant.
> (b) Intensity. This refers to the severity of impact. Responsible officials must bear in mind that more than one agency may make decisions about partial aspects of a major action.[20]

The handbook suggests that the following factors be considered when evaluating the effects of an action:

- Proximity of the action to the targeted resource;
- Timing of the action relative to sensitive phases of the targeted species' life cycle;
- Character of the action with respect to the biology of the targeted species;
- Duration of the effects;
- Frequency of disturbances;
- Intensity of disturbances; and
- Severity of disturbances.[21]

The first four factors generally parallel CEQ's context factor. The last three factors generally parallel CEQ's intensity factor. The factors are interrelated and overlap substantially. For example, the proximity of an action influences its intensity and severity. The timing of an action is a part of its character. Intensity and severity are essentially synonyms. However, the handbook describes intensity in terms of relative effects on population level and severity in terms of effects on recovery rate. In deference to the CEQ guidelines, writers of some EISs and EAs needlessly dwell on separately describing both the context and intensity of each impact. But, teasing out separate discussions of both elements is not essential to good impact assessment and can make impact assessment documents unnecessarily long and complex. Similarly, authors of biological assessments need not dwell on individually considering each factor in the handbook. The important objective is to ensure that the consideration of effects is complete, and accurate, accounting for each factor but not necessarily addressing each factor individually in the text.

Assessing impacts is more than just describing the parameters of an action. Although it is usually necessary to present how many acres of each habitat would be disturbed by an action, an adequate analysis usually has to go further: For example, how might the loss or degradation of the subject area of habitat affect the localized food web? How might the habitat loss fragment migration corridors for specific species? How might established territories for individuals of certain species be disturbed or eliminated? Are the effects temporary or permanent, and if temporary, how completely and how fast are they expected to recover? Ideally, the discussion of impacts should be tied to specific objectives in relevant recovery plans, if available. Even if a recovery plan is not available, the impact discussion should focus on the reasons that the subject species were listed, as outlined in the corresponding *Federal Register* notices.

5.6.5 Impact Assessment from Ecological Risk Assessment Perspective

Although not traditionally applied in either EISs or biological assessments, evaluation of impacts in the context of risk has been advanced conceptually in the last several years. The U.S. EPA defines ecological risk assessment as

> the process that evaluates the likelihood that adverse ecological effects may occur or are occurring as a result of exposure to one or more stressors.[22]

The process views impacts in the context how one or more "stressors" affect ecological "receptors." The process was originally developed to assess the possible effects of chemical contamination on species and habitats at contaminated sites investigated in the context of the Comprehensive Environmental Response, Compensation, and Liability Act (CERCLA; better known as Superfund). In that context, individual contaminants constitute the stressors, and flora and fauna at the contaminated sites constitute the receptors. Specific adverse effects on the receptors are termed "assessment endpoints." Reduced breeding activity and reduced survival of young of a species are possible assessment endpoints. Recognizing that most such assessment endpoints are difficult to quantify, early ecological risk assessors identified "measurement endpoints" constituting easily quantified metrics indicative of possible effects on assessment endpoints. Typical measurement endpoints were developed based on thresholds of response of representative test organisms to gradated exposure concentrations in laboratory settings.

Assessing impacts from a risk perspective is more than just describing the type of impact. It is even more than just quantifying the magnitude of an impact. It is also evaluating the probability of an impact. When assessing the possible environmental impacts from an action, whether in the context of the Endangered Species Act, NEPA, CERCLA, or some other context, one is really predicting the possible results of the action. The action has not yet been implemented. No one knows for sure what the actual effects will be. No one can really state with 100% certainty whether a given possible impact will actually occur. What one can state is that each possible impact has some probability of occurring. Clearly, some impacts have a very high probability of occurrence; habitat cleared of vegetation and paved would almost certainly not be available for use by most species, including most threatened or endangered species, after the action is implemented. But, what about more subtle possible impacts, for example, how wildlife in adjoining natural habitats might respond to noise generated by construction work? It is difficult to generalize how all wildlife would respond. Each species can be expected to respond in its own way, some fleeing from and avoiding any areas receiving the noise, some acclimating to the noise and continuing their former movement patterns, and some responding in an intermediate fashion. Even individuals or groups of individuals of the same species may differ substantially in their response. The body of available scientific knowledge is simply not comprehensive enough to confidently predict how every species, and subsets of those species, living near an action area would respond to a new noise source.

Probability can be mathematically expressed as a number between 0 (impossible) and 1 (certain). Some risk assessment procedures conducted in the context of CERCLA, such as attempts to evaluate possible cancer risks for humans exposed to chemical contaminants, do express risk as a number between 0 and 1. Ecological risk assessment most commonly expresses risk using a hazard quotient (HQ) calculated by dividing contaminant concentration (as a numerator) by some laboratory-derived threshold for significant adverse response (as a denominator). A HQ less than 1.0 is indicative of no potential for significant adverse effects. A HQ equal to or greater than 1.0 is indicative of a potential for significant adverse effects. A HQ is not a probability metric; it is not even a metric proportional to probability. A HQ of 2.0 does not indicate twice the probability of an adverse effect as would a HQ of 1.0, although large HQ differences (on an order of magnitude level) do suggest possibly more severe consequences.

Most impact assessment in the context of NEPA or the Endangered Species Act does not lend itself as readily to quantification, using either direct risk metrics or indirect metrics such as HQs, as does ecological risk assessment in the context of CERCLA. But, that does not mean that evaluations in those contexts cannot at least think conceptually about probability when considering possible impacts. Probability can be expressed at a very conceptual level by using phrases such as "impacts would almost certainly occur," "impacts would likely occur," "impacts could possibly occur," "impacts would be unlikely," or "impacts would almost certainly not occur." Probability can be used as a way of thinking, even if mathematical expression is not possible.

5.6.6 Cumulative Impacts in a Biological Assessment

The need to consider cumulative effects is common to many elements of environmental planning, and the Endangered Species Act is no exception. Environmental impacts do not occur in isolation; the net impact on an environmental resource results from a string of interconnected events of which the proposed action would be just one element. Furthermore, the effective bounds of interrelated actions and their resulting impacts are not usually simple, discrete, and easily defined. Either consciously (purposefully) or subconsciously (inadvertently), agencies can artificially limit the apparent clever way how they define a proposed action. The concept of cumulative impact assessment derives from NEPA. CEQ defines cumulative impacts as

> the impact on the environment which results from the incremental impact of the action when added to other past, present, and reasonably foreseeable future actions regardless of what agency (Federal or non-Federal) person undertakes such other actions. Cumulative impacts can result from individual minor but collectively significant actions taking place over a period of time.[23]

The FWS has developed a similar but not identical definition for purposes of biological opinions under Section 7 of the Endangered Species Act. The FWS defines cumulative impacts as

> those effects of future State or private activities, not including Federal activities, that are reasonably certain to occur within the action area of the Federal action subject to consultation.[24]

There are two key differences. The most obvious is that the Section 7 regulations exclude future unrelated federal actions from the cumulative analysis. The handbook explains that those actions are excluded because the FWS would have to evaluate the impacts from those actions in future action-specific Section 7 consultations.[25] The logic is that the effects of those actions would not be overlooked; they would be considered before the actions are implemented. The second key difference is that the handbook definition encompasses only future actions, while the CEQ definition also notes the contribution of past and present actions—a combination of past, present, and future actions, federal and nonfederal.

Impact assessment purists have reason to consider the Section 7 definition of cumulative impacts at least partially flawed. Cumulative impact assessment strives to understand the environmental impacts of an action in the context of other actions, regardless of their timing or sponsor. Consider a simple case of a proposed allocation of surface water from a reservoir. Commitments of water made in the past, currently under consideration, and reasonably expected to be made in the future, whether to federal or nonfederal recipients, must be considered in order to have an understanding of how much water might remain in the future and what the environmental effects of water withdrawal from the reservoir might have in the future. Failing to account for past commitments or for future commitments to federal users could lead to quantitatively substantial underestimates of overall impacts. The same could be said for other types of incremental impacts, such as habitat loss or losses of individuals of a species.

Although a strict interpretation of the handbook might not require it, the best and most effective biological assessments generally strive to perform cumulative impact assessments that are technically complete, accounting for past, present, and reasonably foreseeable federal and nonfederal actions. They may use the CEQ definition and other guidance issued for the purposes of NEPA. After presenting the complete analysis, they may then add a couple of sentences explaining how the analysis might differ following strict adherence to the handbook definition. Remember, the purpose of a biological assessment is to provide the Services with the information needed to issue any biological opinions and incidental take permits required under the act. Providing a little extra information is not detrimental, as long as the biological assessment does not become excessively lengthy or encyclopedic.

For actions also addressed in an EIS, the cumulative impact assessment used in the biological assessment can draw from that portion of the EIS cumulative impacts text addressing biological resources. It needs to be tailored to the species and habitats covered by the biological assessment. It need not be wordy or complicated; simple and concise yet still technically adequate assessments will be appreciated by the Services. As is true for direct and indirect impacts, tabular presentation of cumulative impacts can be a concise yet effective approach for biological assessments addressing multiple species or habitats.

Cumulative impact assessment is arguably the difficult challenge in environmental impact assessment. Traditionally, many cumulative impact assessments have done little more than restate direct and indirect effects. Cumulative effects must not be confused with indirect effects. Indirect effects are usually addressed together with direct effects, even if cumulative effects are addressed separately. However, an integrated discussion of direct, indirect, and cumulative effects is not only possible but also can even be a more logical and effective approach compared to attempts to parse out any categories of effects. Not all cumulative effects have any connection to the proposed action, other than affecting the same environmental resource.

5.6.7 Biological Assessment Conclusions

Biological assessments commonly present the action agency's conclusions regarding each evaluated species or critical habitat. Many assessments evaluating only one or a few species present conclusions for each species or habitat in separate paragraphs. Assessments evaluating numerous species may present conclusions in tabular format, accompanied by enough text to explain the more complicated or difficult of the conclusions. The handbook outlines a menu of very specific conclusions that the Services should use when responding to informal or formal Section 7 consultations, including

- No effect;
- Is not likely to adversely affect (NLAA);
- Nonconcurrence;[26]
- Is likely to adversely affect (LAA); and
- Is likely to jeopardize.[27]

The act, the regulations, and the handbook do not require action agencies to word their conclusions using these choices, but many action agencies elect to do so. Doing so, however, is a good idea because it helps the Services better understand the action agency's conclusions. The Services will still draw their own conclusions based on their independent review of the biological assessment and other relevant information sources, and the Services' conclusion may differ from the action agency's conclusion.

For a biological assessment, the menu of appropriate conclusions is no effect, is NLAA, and is LAA. The noncurrence and is likely to jeopardize conclusions are typically made by the Services after reviewing the biological assessment. Each of the former three possible conclusions is discussed in further detail as follows:

No Effect: From the perspective of an action agency, a conclusion of no effect is most desirable. If the conclusion is no effect for all listed species and habitats potentially occurring in the action area, there is no requirement for formal consultation. In fact, there is no requirement to submit a biological assessment or even to receive concurrence from the Services. However, many action agencies still seek written concurrence on their no effect conclusions, and some will submit a biological assessment to support their request for concurrence. The handbook provides action agency staff and consultants little guidance on when a no effect conclusion is appropriate. It defines no effect simply as "the appropriate conclusion when the action agency determines its proposed action will not affect a listed species or designated critical habitat."[28]

No effect means just that—no potential effects on the subject species or habitat. It implies a high degree of certainty. Of course, complete certainty is rarely possible; no effect conclusions will be accepted by the Services as long as they are well supported by the best-available scientific evidence even if the possibility of effect cannot be completely ruled out without a shadow of doubt. There is and must be some practicality to the process. If complete certainty (i.e., probability of effect equal to zero) is an asymptotic ideal, one must theoretically accept that some minute probability of effect cannot be ruled out. The Services have not established quantitative thresholds of probability for no effect determinations. Biology and ecology do not always lend themselves to application of quantitative probability theory. But, as a purely theoretical analogy, using the mathematical probability range of 0.0 to 1.0, one may conceptually think of no effect in terms of the lower bound on the range, perhaps less than 0.05 or even 0.01. This evokes confidence intervals of 5% or 1%, which is reflective of those commonly applied in the statistical analysis of experimental results.

Supporting a no effect conclusion can be very challenging for a consultant or other preparer of a biological assessment. The challenge is to prove a negative, the absence of possible effect. Biologists seeking to support a no effect conclusion must theoretically take a weight-of-evidence approach, compiling as much quality scientific evidence as is readily available to support the hypothesis that no effects are possible. A greater weight of evidence is built as more and stronger evidence is provided. Information from multiple sources is generally stronger than information from a single source. At a theoretical level, as the weight of evidence strengthens, the probability of effect moves ever closer to zero. But, the relation is asymptotic, with the probability approaching but never reaching zero. Regardless of how much weight of evidence is accumulated, uncertainty can never be completely eliminated.

It cannot be emphasized enough that the absence of known observations of a species in an action area does not by itself eliminate the possible presence of a species or effects on that species. No one may have ever looked, at least no one who was a qualified observer following a defensible scientific protocol at a proper time of the year. The strongest weight of evidence is offered by one or more targeted surveys of the action area by qualified professionals following survey protocols published or approved by the Services and performed at appropriate times of the growing season immediately preceding implementation of the action. Rarely can this ideal be practicably accomplished. If nothing else, the environmental planning process for most major federal actions requiring biological assessments is usually greater than one year. Hence, the surveys will almost inevitably be performed more than one growing season in advance of the actual disturbances. As most biologists and ecologists recognize, habitats are continually changing, from natural or human-made disturbances as well as natural successional processes.

Hence, the age of surveys or other data must be recognized as a limitation to the evidence provided by those surveys. In general, the older the survey, the weaker is the contribution to the weight of evidence. Agencies dealing with limited staffs and budgets are naturally tempted to rely on dated surveys. And, the Services do not always require updated surveys. They will weigh the age of the surveys against how much uncertainty is presented by relying on dated surveys, attributable to how much they believe the affected habitats may have changed. Of particular importance is the fact that many "comprehensive" surveys of threatened or endangered species for tracts of federal property may be have been performed either before affected species were initially listed or before the Services agreed on appropriate survey protocols.

Some examples of other limitations reducing the contributions by data to the overall weight of evidence, besides the age of the data, include

- Thoroughness of survey effort,
- Qualifications of survey personnel,
- Quality of documentation presenting the data, and
- Statistical support for the data.

Not Likely to Adversely Effect (NLAA): An NLAA conclusion differs from no effect because it acknowledges that effects are possible but have a low probability of being adverse. The handbook states that an NLAA conclusion is "the appropriate conclusion when effects on listed species are expected to be discountable, or insignificant, or completely beneficial."[29]

Beneficial effects result when an action promotes rather than hinders recovery of a species or habitat. Preparers of environmental documents sometimes overlook the possibility that a proposed action might be good rather than bad for the environment. The CEQ recognizes beneficial impacts in the context of NEPA and has established that even beneficial impacts can be significant.[30] Examples of federal actions that might be beneficial for one or more threatened or endangered species include habitat restoration projects, wetland mitigation projects, and projects to control invasive species. Note use of the word *completely* by the FWS when referring to beneficial impacts in the handbook. There are two broad theoretical scenarios where an action might be beneficial in some respects but not completely beneficial:

- The action provides benefits to a listed species but also causes adverse effects to the same species; or
- The action completely benefits one listed species but adversely affects another listed species.

More common than beneficial impacts are adverse impacts that can be considered insignificant or discountable. The handbook states that insignificant impacts "relate to the size of the impact and should never reach the scale where take occurs."[31] Although not developed specifically for use in Section 7 or other elements of the Endangered Species Act, use of CEQ guidance on significance for NEPA[32] can be helpful when writing a biological assessment. As described in Section 5.6.4, the CEQ guidance calls for evaluating the significance of an environmental impact in terms of context and intensity.

An NLAA conclusion is generally not as difficult to support or defend as a no effect conclusion, but it presents many of the same challenges. No effect conclusions are extremely hard to support when there is a distinct possibility that a listed species or habitat occurs in the action area. Instead, NLAA is usually more appropriate when there is a possibility that the species or habitat is present but unlikely to be affected. For example, a biological assessment for a proposed new power plant in central Florida acknowledged the fact that Florida panthers (endangered) might traverse the regional landscape, but that the action lay on the very edge of the known range of the species.[33] Using the probability analogy previously discussed for no effects, an NLAA conclusion would still theoretically imply a probability much closer to zero than 1.0, but perhaps not as tightly tied to the lower end of the range—perhaps 0.05 to 0.1 rather than less than 0.05. This evokes a confidence interval of 5% or 10%. Although confidence intervals greater than 5% are best avoided when performing statistical analysis of experimental results, confidence limits as high as 10% are sometimes used in interpretation of ecological data, where greater variability is common relative to experiments conducted under more controlled conditions.

Action agencies and their consultants commonly propose NLAA conclusions in biological assessments for proposed actions that clearly have a low probability of significant adverse effect, even though they might be able to support no effect conclusions. Whether the Services concur with the NLAA conclusion or instead go with no effect, the project still proceeds in compliance with Section 7 and without conditions or mitigation requirements. The action agency avoids having to defend the more difficult no effect conclusion without risking delay or additional mitigation expenses.

Is Likely to Adversely Affect (LAA): An LAA conclusion is an appropriate conclusion if a no effect or an NLAA conclusion cannot be supported. The FWS defines it as

> the appropriate finding in a biological assessment (or conclusion during informal consultation) if any adverse effect to listed species may occur as a direct or indirect result of the proposed action or its interrelated or interdependent actions, and the effect is not: discountable, insignificant, or beneficial (see definition of "is not likely to adversely affect"). In the event the overall effect of the proposed action is beneficial to the listed species, but is also likely to cause some adverse effects, then the proposed action "is likely to adversely affect" the listed species. If incidental take is anticipated to occur as a result of the proposed action, an "is likely to adversely affect" determination should be made. An "is likely to adversely affect" determination requires the initiation of formal Section 7 consultation.[34]

A useful approach when initially writing a biological assessment is to assume an LAA conclusion for each species and then strive to support a no effect or an NLAA conclusion. If neither of these two lesser conclusions can be supported, then the LAA conclusion remains. It is obviously in the best interest of the action agency for the biological assessment to properly support a no effect or an NLAA conclusion if possible. However, the Services may become suspicious of a biological assessment that asserts the lesser conclusions without adequate support. Good reviewers with the Services will recommend a lesser conclusion for an LAA conclusion in a biological assessment if they are aware of additional information that solidly supports downgrading the conclusion.

Notes

1. This text alludes to similar language used by the CEQ in developing regulatory guidance for NEPA compliance in 40 C.F.R. 1500 et seq., specifically 40 C.F.R. 1500(1)(c), which states "Ultimately, of course, it is not better documents but better decisions that count. NEPA's purpose is not to generate paperwork— even excellent paperwork—but to foster excellent action. The NEPA process is intended to help public officials make decisions that are based on understanding of environmental consequences, and take actions that protect, restore, and enhance the environment. These regulations provide the direction to achieve this purpose."
2. 16 U.S.C. 1536(a)(2).
3. U.S. Department of Energy, Loan Programs Office. About the Loan Programs Office. Available at https://lpo.energy.gov/?page_id=2. Accessed February 23, 2012.
4. Hagerty, T.J. 2005. *Beyond Section 404: Corps Permitting and the National Environmental Policy Act*, Frost Brown Tood, Louisville, KY, August. Available at water.ky.gov/waterquality/Documents/404NEPAand404.doc. Accessed February 4, 2012.
5. 33 U.S.C 1344(c).
6. U.S. Environmental Protection Agency. 2012. Chronology of Section 404(c) Activities, January 30, 2012. Available at http://water.epa.gov/lawsregs/guidance/wetlands/404c.cfm. Accessed February 23, 2012.
7. U.S. Fish and Wildlife Service Endangered Species Act. Home page. Available at http://www.fws.gov/endangered/. Accessed November 2, 2011.
8. NOAA Fisheries Service, Southeast Regional Office, Saint Petersburg, Florida. Available at http://sero.nmfs.noaa.gov/pr/esa/speciesist.htm. Accessed November 2, 2011.
9. Massachusetts Division of Fisheries and Wildlife, Natural Heritage and Endangered Species. Rare Species by Town. Available at http://www.mass.gov/dfwele/dfw/nhesp/species_info/mesa_list/rare_occurrences.htm. Accessed November 2, 2011.
10. U.S. Fish and Wildlife Service. 1998. *Endangered Species Consultation Handbook; Procedures for Conducting Consultation and Conference Activities under Section 7 of the Endangered Species Act*, U.S. Fish and Wildlife Service and National Marine Fisheries Service, Washington, DC. March 1998, Final, p. xv.

11. U.S. Nuclear Regulatory Commission and U.S. Army Corps of Engineers. 2012. *Final Environmental Impact Statement for Combined Licenses (COLs) for Levy Nuclear Plant Units 1 and 2*, Rockville, MD and Jacksonville, MD, in press.

12. U.S. Fish and Wildlife Service. 2006. Guidelines for Bog Turtle Surveys. Revised April 2006. Available at www.fws.gov/northeast/nyfo/es/btsurvey.pdf. Accessed February 28, 2012.

13. Mueller-Dombois, D., and H. Ellenberg. 1974. *Aims and Methods of Vegetation Ecology*, Wiley, New York.

14. U.S. Fish and Wildlife Service, *Endangered Species Consultation Handbook*, p. xi.

15. 10 C.F.R. 51.50.

16. 5 U.S.C. 552.

17. Eccleston, C.H. 2000. *Environmental Impact Statements: A Comprehensive Guide to Project and Strategic Planning*. Wiley, New York, p. 161.

18. U.S. Fish and Wildlife Service, 1998. *Endangered Species Consultation Handbook*, p. 4-25.

19. 40 C.F.R. 1508.8.

20. 40 C.F.R. 1508.27.

21. U.S. Fish and Wildlife Service, *Endangered Species Consultation Handbook*, pp. 4-23 to 4-25.

22. U.S. Environmental Protection Agency. 1998. *Guidelines for Ecological Risk Assessment*, EPA/630/R-95/002F, April, 124 pp. plus appendices. Washington, DC.

23. 40 C.F.R. 1508.7.

24. 50 C.F.R. 402.02.

25. U.S. Fish and Wildlife Service, *Endangered Species Consultation Handbook*, pp. 4-30.

26. Noncurrence is a conclusion that the responding service does not agree with a conclusion presented in a biological assessment or other written document submitted by an action agency under Section 7. In these cases, the responding service usually concludes that the effects fall into one of the other four categories but in disagreement with conclusions presented by the action agency.

27. U.S. Fish and Wildlife Service, *Endangered Species Consultation Handbook*, pp. 3-12 to 3-13.

28. Ibid., p. xvi.

29. Ibid., pp. 3-12.

30. 40 C.F.R. 1508.27(b)(1).

31. U.S. Fish and Wildlife Service, *Endangered Species Consultation Handbook*, pp. 3-12.

32. 40 C.F.R. 1508.27.

33. U.S. Nuclear Regulatory Commission, *Final Environmental Impact Statement*.

34. U.S. Fish and Wildlife Service, *Endangered Species Consultation Handbook*, p. xv.

6

Take Permits and Mitigation

6.1 Introduction

In an ideal world, there would be no take permits. After all, if scientific evidence indicates that a species is in danger of extinction, then a decision to allow individuals to be killed or harmed is, at least theoretically, a decision to contribute to the likelihood of that species' extinction. Nevertheless, we live in a complicated world, and while nearly everyone considers extinction to be a bad thing, some people somewhere do benefit from the activities that cause extinction. Not only do those individuals have constitutional rights to use of private property and "life, liberty, and the pursuit of happiness," but also society as a whole depends on prosperous economic activity that can adversely affect threatened and endangered species and their habitats. The conflict and controversies over private property rights and economic activity versus endangered species is explored in Chapter 8; this chapter focuses on the permitting and mitigation processes under the Endangered Species Act and related environmental statutes.

There are two broad categories of take permits: scientific take permits and incidental take permits. Scientific take permits are generally limited for research and conservation activities. Researchers sometimes need to capture individuals of listed species to propagate them, or in some instances even kill them, to excise tissues and organs for research purposes. As laudable as these activities might be (or at least seem to the researchers), they still require permits. I have not obtained scientific take permits from the U.S. Fish and Wildlife Service (FWS) for federally listed species but have applied for and received similar permits, covering state-listed species, from the Commonwealth of Virginia to collect benthic macroinvertebrates in sediment samples at Dahlgren, Virginia. The purpose of the samples was to collect data to evaluate the development of benthic communities in wetlands restored to mitigate for disturbance necessary to clean up chemical contamination. Whether at the federal or state level, the challenge in the permit application is to convince the issuing agency that the benefits of the research activities to the target species outweigh the loss or harm to the individuals. scientific take permits are not generally controversial.

Practitioners who are not researchers or otherwise involved in not-for-profit conservation activities will probably never apply for or receive a scientific take permit. It is the incidental take permit that might involve most developers or consultants to developers. Incidental take is defined by regulation as "takings that result from, but are not the purpose of, carrying out an otherwise lawful activity conducted by the Federal agency or applicant."[1] Incidental take permits are used where lawful, desirable economic activities such as land development, energy generation, and natural resource extraction conflict with the objectives of the Endangered Species Act. Incidental take permits are the thread that keeps the Endangered Species Act practicable in the real world.

Section 6.2 discusses incidental take permits for federal agencies complying with the Section 7 consultation process. Section 6.3 discusses incidental take permits for private landowners whose actions result in incidental take without the involvement of federal agencies. Section 6.4 is an integrated discussion of mitigation, not only focused on mitigation for threatened and endangered species but also drawing on experience gained from the more common and interrelated wetland mitigation process.

6.2 Incidental Take Permits for Federal Agencies

For federal agencies complying with Section 7 of the act, the formal consultation process culminates in a biological opinion and, if the Services conclude that the action would result in take, an incidental take permit.

6.2.1 Biological Opinions

Section 5.6 describes in detail the biological assessment, the document most commonly prepared by the target audience of this book, environmental consultants and biological staff working for government agencies other than the FWS and National Marine Fisheries Service (NMFS), collectively referred to as the Services. The corresponding document prepared by the Services in response to the biological assessment[2] is the biological opinion. Because this book is not targeted to the staff of the Services, who write biological opinions, the following discussion is less detailed than the corresponding discussion in Section 5.6 for the biological assessment.

The FWS describes the biological opinion as the

> document which includes 1) the opinion of the Fish and Wildlife Service
> or the National Marine Fisheries Service as to whether or not a Federal
> action is likely to jeopardize the continued existence of listed species, or
> result in the destruction or adverse modification of designated critical

habitat; (2) a summary of the information on which the opinion is based; and (3) a detailed discussion of the effects of the action on listed species or designated critical habitat.[3]

The FWS *Endangered Species Act Consultation Handbook*, introduced in the discussion in Chapter 5, presents a suggested outline for biological opinions, consisting of the following parts:

   I. Description of proposed action

  II. Status of the species/critical habitat

 III. Environmental baseline

 IV. Effects of the action

  V. Cumulative effects

 VI. Conclusion

VII. Reasonable and prudent alternatives (as appropriate)[4]

Remember, as noted in Chapter 5, only the Services write biological opinions; action agencies and their consultants write biological assessments. Because the biological assessment functions as an information document provided by the action agency to the Services to facilitate preparation of the biological opinion, action agencies commonly write a biological assessment generally following the same outline recommended for the biological opinion. The Services typically rely heavily on the biological assessment, especially for Part I (description of proposed action) but also in a large part for Part III (environmental baseline), Part IV (effects of the action), and Part V (cumulative effects). However, the Services, not the action agency, author the biological opinion and assume ultimate responsibility for information contained therein, whether obtained from the biological assessment or from other sources.

In particular, Part VI (conclusion) of the biological opinion does not necessarily follow any conclusions presented in the biological assessment. Although action agencies commonly include conclusions in their biological assessments, those conclusions represent the opinions of the action agency, not the Services. Even though the action agencies usually would like the Services to arrive at the same conclusions, the Services would not be performing their duties if they automatically accepted someone else's conclusions. Note that the range of possible conclusions used by the Services is the same as described in Section 5.6 for biological assessments. They include no effect, not likely to adversely affect, and may adversely affect. As in most biological assessments, separate conclusions are presented in biological opinions for each species and critical habitat addressed.

There are two other possible conclusions in a biological opinion. One is "nonconcurrence," which is nothing more than a statement that the Services disagree with the corresponding conclusion presented in the biological opinion. The other is a "jeopardy" conclusion, that is, a conclusion that the action may jeopardize recovery of the affected species. A jeopardy conclusion necessitates that the Services evaluate "reasonable and prudent alternatives," as discussed further in Section 6.2.2.

As with biological assessments, concise, simple, clearly written biological opinions not only will be appreciated by action agencies but also will help the agencies follow the recommendations of the Services. Services personnel should therefore realize that well-written biological opinions further the overall objectives of the Endangered Species Act. Unfortunately, many biological opinions are long, rambling documents with a lot of extraneous text, especially encyclopedic background discussions of species life histories and descriptions of baseline environmental conditions. Such biological opinions further neither the interests of the action agencies nor the objectives of the Services and the Endangered Species Act. Because overworked Service personnel commonly draw extensively from the biological assessment, action agency staff and their consultants can help ensure clearly written biological opinions by clearly writing complete yet concise biological assessments.

6.2.2 Reasonable and Prudent Alternatives

Unlike environmental impact statements (EISs), biological assessments and biological opinions do not serve to compare the effects of alternatives to a proposed action. However, the Services are required to evaluate what are termed "reasonable and prudent alternatives" to the proposed action if (and only if) they conclude in the biological opinion that the proposed action could jeopardize the recovery of one or more listed species (i.e., if the biological opinion reaches a "jeopardy" conclusion). The FWS defines reasonable and prudent alternatives as

> recommended alternative actions identified during formal consultation that can be implemented in a manner consistent with the intended purpose of the action, that can be implemented consistent with the scope of the Federal agency's legal authority and jurisdiction, that are economically and technologically feasible, and that the Director believes would avoid the likelihood of jeopardizing the continued recovery of listed species or the destruction or adverse modification of designated critical habitat.[5]

The concept of "reasonable and prudent" is conceptually reminiscent of the concept of "reasonable" alternatives in the National Environmental Policy Act (NEPA) and the concept of "practicable" used by the U.S. Army Corps of Engineers (USACE) in identification of the least environmentally damaging practicable alternative (LEDPA) when issuing permits under Section 404 of the Clean Water Act. The NEPA statute provides no direction on what constitutes reasonable alternatives to address in the "detailed statement" or EIS.

The Council on Environmental Quality (CEQ) regulatory guidance directs federal agencies to

> rigorously explore and objectively evaluate all reasonable alternatives, and for alternatives which were eliminated from detailed study, briefly discuss the reasons for their having been eliminated.[6]

Specific guidance on what constitutes reasonable alternatives under NEPA is scarce. As with much of NEPA, what constitutes reasonable alternatives is largely left to the discretion of the complying agency and the professional judgment of its staff and consultants. There is no formal enforcement mechanism; excellence in NEPA is largely driven by the possibility of facing lawsuits. What little formal guidance from CEQ regarding alternatives is found in its "Forty Most Asked Questions" document. In its response to Question 2a, CEQ states the following:

> Reasonable alternatives include those that are practical or feasible from the technical and economic standpoint and using common sense, rather than simply desirable from the standpoint of the applicant.[7]

CEQ does not demand scientific purity in EISs, including the selection of alternatives for evaluation. The process is intended to be workable in light of the economic and technological realities confronting agencies and their applicants. Yes, cost is an element in reasonableness; alternatives that might be environmentally elegant but much costlier are not reasonable. The founders of NEPA knew that federal agencies had to be responsible stewards of the taxpayers' money. Especially today, agencies face severe budgetary limitations. However, effective NEPA compliance does not allow for rejecting (or not considering) environmentally preferable alternatives simply because other alternatives are cheaper. Sound judgment is needed to weigh the additional cost of an environmentally preferable alternative against the overall budgetary constraints of the agency.

The USACE must comply with more specific requirements when considering alternatives to proposals requiring permits under Section 404 for filling wetlands and other waters of the United States. The regulatory guidance for evaluating alternatives under Section 404 is sometimes referred to as the "404(b)(1) Guidelines."

> No discharge of dredged or fill material shall be permitted if there is a practicable alternative to the proposed discharge which would have less adverse impact on the aquatic ecosystem, so long as the alternative does not have other significant adverse environmental consequences.[8]

The "aquatic ecosystem" refers to waters of the United States. All surface water features, including wetlands, meeting the regulatory definition[9] are considered part of the aquatic ecosystem, even if they are only infrequently saturated or otherwise more terrestrial than aquatic in character. Although some environmental purists may not like it, a key word in this regulatory language is "practicable." The guidelines provide direction regarding what practicable means:

> An alternative is practicable if it is available and capable of being done after taking into consideration cost, existing technology, and logistics in light of overall project purposes. If it is otherwise a practicable alternative, an area not presently owned by the applicant which could reasonably be obtained, utilized, expanded or managed in order to fulfill the basic purpose of the proposed activity may be considered.
>
> Where the activity associated with a discharge which is proposed for a special aquatic site (as defined in subpart E) does not require access or proximity to or siting within the special aquatic site in question to fulfill its basic purpose (i.e., is not "water dependent"), practicable alternatives that do not involve special aquatic sites are presumed to be available, unless clearly demonstrated otherwise. In addition, where a discharge is proposed for a special aquatic site, all practicable alternatives to the proposed discharge which do not involve a discharge into a special aquatic site are presumed to have less adverse impact on the aquatic ecosystem, unless clearly demonstrated otherwise.[10]

The alternative that minimizes the adverse impact to the aquatic environment while still being practicable is the LEDPA. And, the USACE can issue a Section 404 permit only for the LEDPA. How the USACE reviews an application for a project affecting waters of the United States depends on whether the project is "water dependent." Docks, water intake structures, and dredging navigation channels are examples of water-dependent activities (i.e., they must be placed in or adjacent to surface water to function properly); housing developments, parking lots, and quarries are not. On receiving an application for a non-water-dependent activity encroaching into waters of the United States, the USACE must assume that there is a practicable alternative unless the application demonstrates otherwise. This is sometimes called a "rebuttable presumption"; the burden of disproving this presumption (which is necessary before application can be issued) falls on the applicant.

To close, the process by which the Services must evaluate reasonable and prudent alternatives shares many common themes with alternatives evaluation under other U.S. environmental regulations. These themes are as follows:

- Alternatives must still achieve the general basic objectives of the proposed action.
- Alternatives must be economically reasonable; that is, they may be more expensive than another action for the action agency to implement, but not excessively so.

- Alternatives must be technologically feasible, that is, achievable using existing technology reasonably available to the action agency.

Identifying reasonable alternatives for evaluation under U.S. environmental regulations, whether reasonable and prudent alternatives under the Endangered Species Act, reasonable alternatives under NEPA, or practicable alternatives under Section 404 is very much an art form, relying on the professional judgment of environmental professionals to weigh the interests of the action agency, the public, involved private-sector interests (if any), and the environment. There is no possible cookbook. No generalized discussion can cover the very broad breadth from which alternatives might be appropriate under certain circumstances.

It is possible, however, to identify a few frequently occurring categories of alternatives. These categories generally correspond to concepts of where, how, and when. They include

1. Site alternatives: alternatives that differ with respect to the location of the action, and hence where environmental impacts could occur. These may be thought of as where alternatives.
2. Technology alternatives: alternatives that differ with respect to what techniques or procedures are implemented at a given site to achieve an objective. These may be thought of as how alternatives.
3. Timing alternatives: alternatives that differ with respect to when an action is implemented. These may be thought of as when alternatives.

Site alternatives are perhaps the most obvious. If threatened or endangered species, or other protected resources under regulations other than the Endangered Species Act, occur at a proposed site, consider moving the project to a different location lacking these resources, or at least fewer or less-valuable resources. Of course, many actions are simply not amenable to being redirected to another location. If the proposed action is that of a private-sector applicant involving privately owned land, making an applicant move an action to another site can be construed as encroaching on constitutionally protected private property rights.

Possible technology alternatives depend on the type of proposed action. For a power plant, some possible technology alternatives might be alternate power generation technologies such as coal, natural gas, nuclear, solar, wind, or geothermal energy. For a public transportation project, some possible technology alternatives might include a new highway, upgrading existing highways, or a light rail connection. For a timber management project, possible technology alternatives might include clear-cutting, selective cutting, or seed tree regeneration. For a housing development, different technological approaches might include a grid of single-family lots, clustered singe-family lots, clustered townhouse lots, or clustered multifamily units.

Timing alternatives can be especially useful in the context of the Endangered Species Act. Moving habitat-disturbing activities to periods when sensitive species are not present can be highly effective in avoiding impacts to those species. Many migratory birds are present in parts of North America only during their breeding season, usually between March and October for most species in most areas. Some species, such as the endangered Indiana bat (*Myotis sodalis*), hibernate in caves or other areas in the winter; actions away from the caves or other hibernating sites (hibernacula) can be conducted without physically disturbing those species during known hibernating seasons.

The preceding compartmentalized discussion should not be interpreted as encouraging a formulaic approach to identifying alternatives. Not all possible alternatives fall into just one of the where, how, or when categories discussed.

6.2.3 Incidental Take Statements

If the biological opinion includes a conclusion of may adversely affect for one or more listed species or critical habitats, the Services must also issue an incidental take statement. The requirement for an incidental take statement exists whether or not there is a jeopardy determination warranting an evaluation of reasonable and prudent alternatives—as long as the Services' conclusions for one or more species or critical habitats in the biological opinion is may adversely affect. The statutory language establishing the incidental take statement states:

> The Secretary shall provide the Federal agency and the applicant concerned, if any, with a written statement that
>
>   i. specifies the impact of such incidental taking on the species,
>   ii. specifies those reasonable and prudent measures that the Secretary considers necessary or appropriate to minimize such impact,
> iii. in the case of marine mammals, specifies those measures that are necessary to comply with section 1371 (a)(5) of this title with regard to such taking, and
> iv. sets forth the terms and conditions (including, but not limited to, reporting requirements) that must be complied with by the Federal agency or applicant (if any), or both, to implement the measures specified under clauses (ii) and (iii).[11]

The term *federal agency* in this above language refers to the action agency. The reference to the *applicant* refers to any nonfederal parties seeking permits or other authorizations from federal agencies forced to comply with Section 7 of the act. If the action is exclusively that of a federal agency without the involvement of nonfederal applicants, then any reference to applicants may be ignored. The FWS has established that the contents of an incidental take statement should include the following:

- Introductory paragraph
- Amount or extent of take anticipated
- Effect of the take (to comply with Item i)
- Reasonable and prudent measures (as appropriate) (to comply with Items ii and iii)
- Terms and conditions (to comply with Item iv above)
- Coordination with other laws, regulations, and policies[12]

From the perspective of the action agency, there are two key elements of an incidental take statement. The first element is quantification of the amount of take authorized. That limit is typically expressed as a number of individuals or area of habitat that can be killed or disturbed without violating the Endangered Species Act.[13] The Services do not issue blank checks authorizing unlimited or indeterminate impacts to listed resources; the limits expressed in each incidental take statement are carefully developed based on careful review of the best-available scientific information. Action agencies are responsible for ensuring that their actions do not exceed the maximum take authorized in the incidental take statement.

The second element establishes the reasonable and prudent measures that must be taken by the action agency to minimize the effects of the take and terms and conditions that must be observed by the recipient in carrying out the proposed action. Reasonable and prudent measures are defined by the FWS as "actions the Director believes necessary or appropriate to minimize the impacts, i.e., amount or extent, of incidental take."[14] The act limits what the Services can require from action agencies; the Services cannot extort concessions from other federal agencies (or their applicants) that are not justified under the statutory language of the act. The FWS considers measures to be reasonable and prudent if they are consistent with the general design, location, scope, duration, and timing of the proposed action.[15]

Reasonable and prudent measures are different from reasonable and prudent alternatives, but both do share the concept of "reasonable and prudent." Reasonable and prudent alternatives are substantial variations in the design of an action, such as selecting a different site or technology, to accomplish a stated objective while reducing or eliminating potential impacts to protected species or habitats. Reasonable and prudent measures are minor changes or minor follow-up activities to a selected alternative that reduce the potential for impacts. The former may be thought of as changing a channel; the latter may be thought of as fine-tuning. Some examples of possible reasonable and prudent measures might include

- Adjusting the proposed footprint of construction to exclude or work around inclusions of habitat on a site that are of particular importance to listed species;

- Preserving buffers of forest on steep slopes, in floodplains, or along the shores of streams or other surface water bodies;
- Establishing nest boxes or artificial roosts for birds;
- Reducing noise levels in certain portions of a site or at certain times (e.g., expected nesting seasons);
- Establishing brush piles or other general habitat improvement features;
- Adding colored balls or other objects to transmission lines to increase visibility to birds, reducing the probability of collisions;
- Spacing transmission line conductors in a manner that reduces the probability of birds touching two or more conductors (phases) simultaneously, resulting in electrocution;
- Placing mats or nets over wetlands or erodable soils when temporarily encroached on by heavy equipment;
- Relocating individuals of listed wildlife species, or transplanting individuals of listed plant species, to other areas of suitable habitat;
- Erecting road signs warning drivers to slow down and drive especially carefully in areas with a high likelihood of having listed species;
- Building passages allowing movement of wildlife under highways;
- Erecting fences or other barriers to keep construction equipment and workers from entering areas of sensitive habitat;
- Erecting silt fences and other temporary erosion control structures during facility construction; and
- Seeding temporarily disturbed areas of habitat with regionally indigenous plants and planting seedlings of regionally indigenous trees and shrubs.

Again, this list only provides examples. Other measures might be reasonable and prudent for certain proposed actions. Some of the measures noted in the list might not be reasonable or prudent for some proposed actions. For example, restoring temporarily disturbed areas might not be reasonable with respect to cost in urban areas, where using the land for conservation might require forfeiting lucrative urban uses. Building wildlife passages could be cost prohibitive, depending on the unique engineering constraints of a particular highway stretch. Or, the cost might not be warranted unless a high level of use by targeted wildlife can be expected. Seeding or planting tree saplings might not be reasonable if easily accessible water sources are not available. The labor needed to establish and maintain tree and shrub plantings might be warranted for large "big ticket" projects but not for smaller, lower-budget actions.

6.3 Incidental Take Permits for Nonfederal Applicants

Nonfederal entities conducting lawful activities adversely affecting listed species or habitats do not have to follow the Section 7 consultation process but still must obtain incidental take permits from the services under the Endangered Species Act. As emphasized repeatedly in this book, the fact that a proposed action is sponsored by a private developer or other nonfederal party does not necessarily mean that it escapes the Section 7 process; if the party must obtain a permit, authorization, or funding from a federal agency, that agency must follow the Section 7 process. If so, the agency indirectly involves the nonfederal party in the Section 7 process. The common federalizing trigger bringing private developers into the Section 7 process is the requirement to obtain Section 404 permits from the USACE for impacts to wetlands or other surface water features.

But, not all private development projects require Section 404 permits or other federal support or authorization. Smart developers can often design their project to avoid wetlands and other areas regulated under Section 404. Agricultural, ranching, and forestry activities, which are generally exempt from Section 404, are sometimes capable of adversely affecting threatened or endangered species.

6.3.1 The Permit Application Process

Nonfederal interests proposing development activities adversely affecting listed species or critical habitats follow a process not unlike the Section 7 process. The regulations established in 50 C.F.R. 17.22 state that an application must include

- A complete description of the activity sought to be authorized;
- The common and scientific names of the species sought to be covered by the permit, as well as the number, age, and sex of such species, if known; and
- A conservation plan that specifies
  - The impact that will likely result from such taking;
  - What steps the applicant will take to monitor, minimize, and mitigate such impacts, the funding that will be available to implement such steps, and the procedures to be used to deal with unforeseen circumstances;
  - What alternative actions to such taking the applicant considered and the reasons why such alternatives are not proposed to be utilized; and
  - Such other measures that the director may require as being necessary or appropriate for purposes of the plan.[16]

Even casual reading of these regulatory requirements indicates that what the applicant must submit is a package that contains many of the substantive elements of a biological assessment, as described for the Section 7 process. Like a biological assessment, the applicant must include a detailed description of the proposed project, an indication of what listed species or habitats are subject to impacts, and a description of the possible effects of the action. Most of what is presented in Chapter 5 on these elements of a biological assessment applies here as well. Of particular importance is the need to provide as detailed and accurate description of the proposed project as possible; even if Services personnel have the expertise to fill in gaps regarding the description of impacts, they must depend on the applicant and the applicant's consultants to describe what the applicant wants to do. Vague, incomplete, or inaccurate descriptions will inevitably result in delays or complications in the Endangered Species Act compliance process and could even lead to violations. Any costs originating from incomplete or inaccurate project descriptions cannot be attributed to the Service or the act—they lie completely with the applicant.

The "conservation plan" mentioned in the regulations is the habitat conservation plan. The plan requires two elements not normally included in a biological assessment, but both are elements that could ultimately be required for federal actions in the Section 7 process, depending on the severity of possible impacts. The first is a statement of what steps will be taken to monitor, minimize, and mitigate impacts. This is the heart of a habitat conservation plan. When the Endangered Species Act was first implemented in the 1970s, it contained no provisions for authorizing take of listed species, except as necessary for scientific research or conservation activities. Take was to be avoided except in unusual circumstances involving research and conservation actions; development and other economic land use activities had to avoid take—no exceptions. The current system allowing for incidental take permits for federal or nonfederal projects constitutes a compromise between environmentalists and economic interests in the early years of the Reagan administration. This compromise represented an inevitable stage of maturation for the Endangered Species Act; without it, the act would likely have been repealed, perhaps rightly so.

The second element required in a habitat conservation plan is an evaluation of possible alternative actions. As stated previously in Section 6.2, alternatives analyses are a common requirement under U.S. environmental regulations, and the Endangered Species Act is no exception. Under Section 7, an evaluation of "reasonable and prudent alternatives" is required if the federal action contained a jeopardy conclusion—that the action could jeopardize recovery of one or more listed species. All habitat conservation plans must consider alternatives. The detail needed for the alternatives analysis is dependent on the complexity of the project and severity of potential impacts. The discussion of alternatives analysis included in Section 6.2.2 for Section 7 is useful in the development of habitat conservation plans as well.

6.3.2 Habitat Conservation Plans

The FWS has prepared a handbook on preparing habitat conservation plans.[17] The handbook describes the permitting process in detail but provides few specifics or technical direction in preparing plans. Instead, the handbook emphasizes close communication between the applicant and the Services throughout plan development. This is a common theme permeating the processes for complying with the Endangered Species Act; the FWS emphasizes the need for close coordination between federal action agencies and the Services in the Section 7 consultation process. As with federal agencies, the Services typically have a greater depth of specialized biological expertise than do either other federal agencies or most landowners or corporate interests.

The development of a habitat conservation plan is commonly described as a "partnership" between the applicant, usually a private-sector developer or landowner, and the Services. In this partnership, the applicant receives permission for limited take unavoidable resulting from economic use of private property in exchange for promising to perform conservation-oriented land management practices that benefit the recovery of listed species, hence furthering the objectives of the Endangered Species Act.

Regional versus Individual Habitat Conservation Plans: Most of the earlier habitat conservation plans were relatively simple documents addressing proposed work on a single tract of nonfederal property. While these simple habitat conservation plans continue to be common, many habitat conservation plans have been submitted and approved for blocks of hundreds of thousands of acres of property, sometimes owned by multiple property owners. These landscape-level documents can help to simplify the process by instituting conservation practices over large areas, avoiding the need for multiple individual landowners preparing similar plans for multiple tracts of property containing similar habitats for the same listed species. According to the FWS, "Habitat conservation plans are evolving from a process adopted primarily to address single developments to a broad-based, landscape level planning tool utilized to achieve long-term biological and regulatory goals."[18]

The increased use of regional habitat conservation plans parallels two other similar trends toward consolidating and simplifying compliance with U.S. environmental protection laws. First, it parallels the increasing use of tiering programmatic environmental impact statements (PEISs) in NEPA. Section 1502.20 of the CEQ guidance regulations states:

> Agencies are encouraged to tier their environmental impact statements to eliminate repetitive discussions of the same issues and to focus on the actual issues ripe for decision at each level of environmental review (Sec. 1508.28). Whenever a broad environmental impact statement has been prepared (such as a program or policy statement) and a subsequent statement or environmental assessment is then prepared on an action included within the entire program or policy (such as a site specific action) the subsequent statement or environmental assessment

need only summarize the issues discussed in the broader statement and incorporate discussions from the broader statement by reference and shall concentrate on the issues specific to the subsequent action.[19]

Although tiering does not eliminate the need for NEPA compliance documentation for lower-level actions, it does avoid the need for repeated evaluation of high-level issues in multiple lower-level EISs. Done properly, tiering can serve to simplify lower-level EISs as well as sometimes allow for the use of environmental assessments (EAs) and findings of no significant impact (FONSIs) in lieu of EISs for lower-level actions.

Second, the use of regional habitat conservation plans parallels the use of general permits under Section 404 of the Clean Water Act.[20] The USACE issues groups of general permits granting blanket authorization of clearly defined categories of actions falling within the parameters of the permits. Although a few general permits covering very simple activities with very small potential impacts to regulated waters allow for project proponents to carry out activities qualifying under the general permit without notifying the USACE, most general permits require proponents to submit notification documents to USACE demonstrating that the proposed action falls within the scope of the general permit.[21]

Habitat Conservation Plans: The core element of a habitat conservation plan is a proposal of mitigation measures that avoid, minimize, or compensate for impacts to listed species. A unified discussion of mitigation applicable not only to habitat conservation plans but also to the Section 7 consultation process as well as to related U.S. environmental laws such as NEPA and Section 404 is presented in Section 6.4. Mitigation measures are perhaps the most tangible and visible contributions of environmental compliance efforts to the environment. Their value transcends what sometimes seems to be an abstract paperwork exercise. Mitigation measures do, however, usually come with costs and project delays that can extend beyond the initial planning and permitting phases of a project. Preparers of habitat conservation plans should therefore propose mitigation measures very carefully and only after close and frequent communication with the Services. While the objective is to propose adequate but not excessive mitigation, the process of developing mitigation measures can involve back-and-forth "horse-trading" between an applicant seeking to minimize project costs and delays and the Services, seeking to further the overall objectives of the Endangered Species Act. Mitigation measures should be developed in an extended process of frequent communication between the applicant, the applicant's consultants, and the Services rather than presented in a single grandiose document.

No Surprises Rule: A controversial compromise between the Clinton administration and private property activists implemented in 1998 is sometimes referred to as the "no surprises rule." The no surprises rule[22] assures private landowners receiving an incidental take permit that the Services will not hold permittees liable for future, unanticipated ecological consequences in response to unanticipated ecological consequences related to the species covered by the permit. The Services would agree to limit mitigation requirements to those outlined in the approved habitat conservation plan and not impose additional mitigation measures for the same species. The rule provided property owners willing to work with the Services through the established incidental take permit process with a level of surety that additional future mitigation measures would not be imposed in response to unforeseen circumstances arising with respect to the subject species. One of the most frequent complaints of business interests and property owners is that complying with U.S. environmental laws can lead to an unpredictable, open-ended sequence of follow-up mitigation requirements. Unpredictable mitigation requirements are an especially difficult challenge to business planning and budgeting. The no surprises rule is a commonsense compromise that has helped to make the Endangered Species Act workable in the real world.

Safe Harbor Agreements: Another commonsense compromise related to the Endangered Species Act instituted in the late 1990s by the Clinton administration is the provision for safe harbor agreements.[23] A property owner wishing to perform conservation measures on private property that potentially benefit listed species may propose a safe harbor agreement under which the Services authorize future take of the subject species at levels that do not exceed the expected benefit from the conservation measures. The provision for safe harbor agreements was implemented in response to landowners desiring to implement conservation measures but had been deterred because increased numbers of protected species could limit future economic use of the property. One of the first safe harbor agreements developed under the provision was proposed by a group of agricultural property owners in the Lower Mokelumne River Watershed in San Joaquin County, California.[24] The owners wanted to establish riparian vegetation along streamsides in the watershed that could be attractive to the valley elderberry longhorn beetle (*Desmocerus californicus dimorphus*), listed as threatened. In particular, the beetle favors riparian vegetation containing native elderberry (*Sambucus* spp.) shrubs. The safe harbor agreement provides assurances to the landowners that their voluntary proposal to establish riparian vegetation would not limit future uses of the property, at least with respect to the valley elderberry longhorn beetle.

NEPA Compliance for Habitat Conservation Plans: Although nonfederal applicants applying to the Services for incidental take permits are not subject to NEPA, the Services are themselves federal agencies whose actions are subject to NEPA. If the act of issuing an incidental take permit results in significant impacts to the environment, the Services must prepare an EIS prior to issuing the permit. The Services recognize three scenarios under which an incidental take permit associated with a habitat conservation plan can be issued:

1. Low-effects habitat conservation plan, a simple document proposing only very minor impacts, which can be approved expeditiously by the Services using a categorical exclusion;
2. Habitat conservation plan proposing more extensive impacts that, with adequate mitigation, would not result in significant environmental impacts and can be issued with an EA; and
3. Habitat conservation plan proposing more extensive impacts that even with mitigation could be significant or controversial and must be issued with an EIS.[25]

Because habitat conservation plans include mitigation, few require an EIS for the Services to approve. If the Services must prepare an EA (or less likely, an EIS), they usually seek extensive input from the applicant. Even though the conclusions of the EA (or EIS) are those of the Services, the applicant will often bear the burden of performing much of the data collection legwork.

6.4 Mitigation

The concepts of reasonable and prudent alternatives and reasonable and prudent measures encompass substantial elements of a key concept inherent to many U.S. environmental regulations: mitigation. Simply stated, mitigation is a sequential planning process that seeks to identify and avoid possible adverse environmental impacts, minimize the effects of those impacts that cannot be completely avoided, and implement measures to offset the adverse effects of unavoidable impacts. Even more simply stated, mitigation is an effort to predict possible impacts, avert as many of those impacts as possible, and attempt to rectify for those that cannot be averted. Mitigation is not just a single action; it is an extended planning process. It first entails avoidance, with compensation as a last resort. Most of the misunderstandings and controversies surrounding mitigation ultimately derive from thinking of mitigation as a ready tool for impact compensation, without viewing that compensation as the end product of a screening process involving the identification of impacts and evaluation of the practicality of avoiding those impacts.

In developing its guidance for implementing NEPA (summarized in more detail in Chapter 4 of this book), the CEQ defined *mitigation* as

    a) Avoiding the impact altogether by not taking a certain action or parts of an action.
    b) Minimizing impacts by limiting the degree or magnitude of the action and its implementation.
    c) Rectifying the impact by repairing, rehabilitating, or restoring the affected environment.

d) Reducing or eliminating the impact over time by preservation and maintenance operations during the life of the action.

e) Compensating for the impact by replacing or providing substitute resources or environments.[26]

Mitigation is arguably the most important, and certainly the most tangible, element of environmental planning and regulation. Identifying and evaluating possible effects on environmental resources, including threatened and endangered species, is a necessary preliminary step. But, it ultimately serves to provide a technical basis for mitigation. If paying attention to the road while driving is important (and no one can argue that it is not), taking action to miss hitting a pedestrian or keeping from driving off a cliff is the even more important follow-on measure. Much of the public view environmental planning and regulation as little more than a procedural hurdle, "jumping through hoops," dotting some *i*'s before the bulldozers can be finally cranked. From this arises the perception that environmental planning is a waste of time, money, and resources. At best, they might view the process as a weak winnowing process, perhaps screening out a few half-hearted development proposals not backed by the developer's determination and purpose, maybe something along the often-quoted words surrounding NEPA: "look before you leap." Only through the more demonstrative, "concrete" (no pun intended), manifestation of avoidance accomplishments and visible compensatory efforts—those plainly visible nature preserves with their interpretive trails and fascinating signage—do many nonexperts plainly see the benefits, the fruits of the process. Averting a few emotionally damaging projects (i.e., that massive new international airport planned for the Everglades in the 1960s) may have carried the freight for initially establishing many of the environmental regulations we take for granted today, but if the environmental history of the three decades since 1980 proves anything, simple emotion will not serve to carry these regulations indefinitely. The public must be convinced of the successes of these regulations, and that means visibly effective mitigation.

6.4.1 Mitigation under the National Environmental Policy Act

Most EISs include a mitigation measure section addressing its proposed action. Theoretically, proposing mitigation in the context of NEPA is optional, as NEPA is primarily a mechanism for disclosing possible environmental impacts and deciding among possible alternatives. However, agencies know that proposing inadequate mitigation could prompt lawsuits claiming that the agency did not meet the spirit and intent of NEPA. Furthermore, agencies responsible for issuing permits or formal consultations required for the proposed action may withhold those approvals until adequate mitigation is proposed. This includes possible refusals by one or both Services to consult formally under Section 7 of the Endangered Species Act. Once the final EIS is finished, agencies typically commit to implementing the proposed mitigation measures in their record of decision (ROD) for the project. One agency, the U.S. Department of Energy, prepares a mitigation action plan (MAP) outlining how mitigation presented in an EIS will actually be implemented.[27] Whether presented in a ROD or a MAP, agencies advance the objectives of NEPA best when they clearly present the specifics for all proposed mitigation.

Mitigation discussions usually sound good when presented in an EIS. It is not uncommon for an EIS to describe elaborate mitigation measures and provide detailed scientific justifications regarding how the mitigation would offset the impacts. But, most EISs are long and complicated documents, and few outside of specialists and regulatory personnel read them in detail. But, as interesting and well written these descriptions of proposed mitigation might be, they are of absolutely no value to the environment if they are never implemented or never implemented wholeheartedly. Producing follow-on documents such as MAPs is one way to help ensure follow-through on proposed mitigation. Still, the NEPA process can be an excellent framework for development of environmental mitigation as long as the action agency is held responsible for successful implementation of the mitigation proposed in the EIS.

Until recently, EAs did not commonly propose mitigation measures because EAs were completed only for actions lacking no potentially significant environmental impacts (i.e., actions that qualified for a FONSI). The rationale was that an action resulting in significant environmental impacts warranted an EIS, with its greater opportunities for public involvement, even if the action agency planned to mitigate for those impacts. CEQ's response to Question 40 in its "Forty Most Asked Questions" stated:

> Mitigation measures may be relied upon to make a finding of no significant impact only if they are imposed by statute or regulation, or submitted by an applicant or agency as part of the original proposal. As a general rule, the regulations contemplate that agencies should use a broad approach in defining significance and should not rely on the possibility of mitigation as an excuse to avoid the EIS requirement.[28]

However, agencies have recently prepared numerous EAs for actions that qualify for FONSIs conditioned on mitigation. While CEQ and EPA once discouraged such an approach, the use of what is now termed a "mitigated FONSI" is no longer discouraged. A task force established in 2003 to modernize NEPA practice recommended that CEQ establish more detailed guidance on the use of mitigation to qualify actions for FONSIs.[29] The current thinking is that if the net environmental impact of an action and its associated mitigation is minimal, then there is no reason that it cannot be handled by the same process used for other actions with minimal impacts. Whether one agrees or disagrees with the use of mitigated FONSIs is largely a reflection of one's confidence in the capabilities of mitigation.

6.4.2 Wetland Mitigation

The U.S. EPA and USACE have developed a similar approach to mitigation in the context of Section 404 of the Clean Water Act following a sequential consideration of avoidance, minimization, and compensatory mitigation. Section 404 mitigation is commonly termed "wetland mitigation" even though Section 404 regulates not only wetlands but also most other surface water bodies, termed "waters of the United States" (see Section 4.3 of this book). Wetland mitigation places a strong emphasis on avoiding and minimizing encroachment into wetlands (and other waters of the United States) before even considering compensatory mitigation. Until 2008, the USACE and EPA encouraged "on-site, in-kind" compensatory mitigation comprising efforts to enhance, restore, or create wetlands on the project site. The rationale was that compensatory wetland projects establishing new wetlands as identical as possible to the impacted wetlands would benefit the same location (service the same receptors) in the same way that the lost wetlands formerly had. Often, however, on-site, in-kind wetland mitigation was not practicable, usually due to the limited area or unfavorable physical characteristics on the site. In such cases, the agencies would accept off-site wetland mitigation projects, although they would still encourage the off-site mitigation to be as in kind as possible and to be sited as close as possible to the site. Restoration of former wetlands that had been degraded by previous activity was encouraged because such projects, typically involving plugging drainage ditches or otherwise restoring previous hydrological conditions, had a greater probability of success than trying to establish wetlands where they had not previously occurred. Enhancement of existing wetlands was also encouraged. However, compensatory payments to agencies to help fund wetland mitigation activities (sometimes called "payments-in-lieu") was usually encouraged only as a last resort. Such payments, while usually favored by permittees, placed the risk of mitigation project failure on the agencies rather than on the applicant.

The United States is an entrepreneurial country. Once wetland mitigation became well established and substantial demand existed for off-site wetland mitigation, private enterprises began to establish wetland conservation projects voluntarily and sell interests in the projects to Section 404 permittees confronted with having to identify off-site wetland mitigation opportunities. Permittees required to perform a specified acreage of wetland mitigation (and who could not practicably perform the mitigation on site) could purchase the requisite acreage from the privately sponsored project instead of completing their own off-site project. The process was termed *mitigation banking*, and the projects were termed *mitigation banks*. The USACE and EPA were initially reluctant to embrace mitigation banking but eventually warmed to the concept for cases where on-site mitigation was not practicable. Certain limits, however, were usually placed on the use of mitigation banks. The portion of the mitigation bank purchased by the permittee usually had to be at least partially in kind to the impacted wetlands. The bank had to be within the same watershed (at least at a high level) and in the same general physiographic setting as the project site.

I spent part of the late 1990s and early 2000s designing wetland mitigation projects for hazardous waste cleanup sites on a Navy base in northern Virginia. The base went through multiple name changes over that period but can be for simplicity referred to as the Naval Surface Warfare Center Dahlgren, or Dahlgren for short, in Dahlgren, King George County, Virginia. Most of the sites contained contaminated surface soils or sediments[30] that had to be removed or capped to protect human health and the environment. Because part of the contamination at each site was in wetlands, complete wetland avoidance was not possible. A remedial investigation and feasibility study (RI/FS) completed under the Comprehensive Environmental Response, Compensation, and Liability Act (CERCLA, sometimes referred to as Superfund) had concluded that some form of active remediation was necessary to protect human health and the environment and hence that simply leaving the wetlands untouched was not practicable. The remedies were designed to limit nearly all wetland disturbance to those wetlands containing contaminated soil or sediment, keeping project elements such as construction equipment staging areas outside uncontaminated wetlands, thereby minimizing wetland impacts. The remedial designs also called for implementing best management practices (BMPs) to minimize soil erosion and sedimentation into the wetlands and storm water management practices to minimize erosion and other hydrological damage to the wetlands by storm runoff.

Finally, the planning effort culminated in the design of compensatory mitigation. The logical compensatory mitigation at most of the sites was simple: restore the contaminated wetlands damaged by removal of tainted soil and sediment to their original physical properties using clean soil and sediment. This constituted a type of compensatory wetland restoration. It did not involve creating wetlands where they had not formerly existed (wetland creation). It really did not involve enhancing the contaminated wetlands; while removing the contamination could be described as a type of enhancement, the excavation process essentially destroyed the affected wetland areas, leaving something requiring restoration rather than mere enhancement. Last, the mitigation could easily be described as on site and in kind; the expectation of the designs was that restoring existing water depths, tidal flow regimes, and runoff and site flow conditions would reestablish the conditions necessary to regenerate the same kind of vegetation. I sometimes referred to it as "in situ, in-kind" wetland restoration, a stronger version of the idealized on-site, in-kind wetland restoration.

An especially creative wetland mitigation project is Freedom Park in the city of Naples, Florida. Freedom Park contains a constructed wetland system established to offset impacts to wetlands resulting from widening and upgrading Airport-Franke Road, an arterial highway in Naples. As the area used to construct the new Freedom Park wetland system comprised abandoned farmland, the new wetlands can be considered an example of wetland creation. The new wetlands were designed to capture runoff from the newly widened highway, allow sediments and oils and greases to settle out of the detained runoff water, and convey the cleaned water to natural wetlands bordering the Gordon River, a tidal waterway traversing much of Naples. The sequence of shallow pools bordered by emergent wetlands conveys runoff through multiple detention stages, capturing sediment and conveying the water to a natural cypress swamp bordering the Gordon River. Invasive plants such as Brazilian pepper were removed from the cypress swamp, and native tree saplings were planted, adding elements of wetland enhancement and preservation to the project. To top it off, the city constructed walking paths around the constructed pools and boardwalks through the preserved and enhanced cypress swamp, providing educational and recreation opportunities to nearby schools, residents, and tourist destinations.

The traditional framework for wetland mitigation is shown in Table 6.1. However, that framework was to be turned on its head in 2008. In that year, EPA and USACE issued what is termed the "mitigation rule" outlining a new approach to wetland mitigation under Section 404.[31] The mitigation rule changed the language of USACE's and EPA's Section 404 regulations to address long-standing concerns over the perceived success of wetland mitigation projects implemented over the pervious 20 years. Many within the permitting agencies felt that wetland mitigation projects would have a greater probability of success if conducted by experienced experts employed by natural resource agencies or nonprofit organizations or firms specializing in wetland mitigation. The mitigation rule therefore encourages permit applicants first to consider the availability of suitable wetland mitigation banks before proposing permittee-sponsored mitigation, even if on site, in kind. The mitigation rule also encourages permittees to consider payments-in-lieu before proposing their own mitigation projects. Once these two approaches are considered, the mitigation rule allows for permittee-sponsored mitigation in general accordance with the preferences observed prior to the rule. Although the mitigation rule has established mitigation banking and payments-in-lieu as mitigation preferences, individualized permittee-sponsored mitigation has still been allowed, especially for large projects in geographic areas not serviced by appropriate mitigation banks. Some states with state-level wetland permitting and mitigation requirements also still encourage on-site, in-kind mitigation. The mitigation rule is still too new to evaluate what its ultimate effect on wetland mitigation trends will be.

Quantifying wetland mitigation requirements was traditionally based on the theory of "no net loss" of wetlands, based on wetland acreage. The minimum was a 1:1 mitigation ratio, that is, establishing one acre of wetland for each acre of wetland lost. Most USACE districts established higher mitigation ratios for forested wetlands, recognizing that newly created or restored wetlands would be planted with tree saplings or seedlings that would require several years to grow into a mature forest and fully functioning

TABLE 6.1

Mitigation Sequencing under Section 404 of the Clean Water Act

Approach	Subapproach	Description	Examples	Advantages and Disadvantages

Avoidance		Designing project not to impact wetlands (or impact as few wetlands as practicable)	Rerouting a proposed highway around wetlands
			Clustering lots so wetlands on project site can be preserved
				Pro:
				  Certainty
				  Generally less costly
				Con:
				  Limits flexibility

Minimization		Taking actions to reduce impacts to wetlands that cannot be avoided	Implementing best management practices to reduce sedimentation and runoff into wetlands
			Placing nets or pads over wetlands before entry by vehicles
				Pro:
				  Certainty
				  Generally less costly
				Con:
				  Limits flexibility

Compensatory mitigation	Wetland restoration	Converting former wetlands back into wetlands	Plugging drainage ditches Excavating and removing fill	Pro: Proper landscape position and known history of success at site Con:

	Wetland creation	Establishing wetlands where they did not formerly occur	Excavating uplands to establish ground surface close to water table
			Diverting runoff into dry depressions
				Pro:
				  Acreage and functional gain
				Con:
				  Uncertain success
				  Generally more expensive

			Pro/Con
Wetland enhancement	Improving functionality of existing wetlands	Removing exotic or invasive vegetation from wetlands. Installing nest boxes in forested wetlands	Pro: Functional gain Con: No increase in wetland acreage
Mitigation banking	Interest (credits) sold to developers by third-party entrepreneurs performing off-site mitigation	Any of several compensatory procedures (restoration, creation, enhancement, or preservation) can be used to establish bank	Pro: Shifts risk to conservation specialists. Favored by mitigation rule Con: Credits can be costly
Wetland preservation	Wetlands purchased or set aside for protection from economic activity	Buying and donating private land to public conservation uses	Pro: Relative certainty. Not costly in areas with cheap land Con: Costly in urban areas. No increase in wetland acreage or function
Payment-in-lieu	One-time payment of a negotiated sum of money to conservation agency	Donating money to agency or nonprofit entity for conservation purposes	Pro: Shifts uncertainty burden to agencies. Encouraged by mitigation rule when bank credits not available Con: Funds can be abused

forested wetland. Some common ratios were 2:1 (two acres of new wetland for each acre of wetland lost) for forested wetlands and 1.5:1 for scrub-shrub wetlands. The minimum ratios were common for mitigation in the form of wetland creation or restoration, but even higher ratios were usually required for enhancement, recognizing that the enhanced areas still possessed some wetland qualities even before enhancement. The USACE often established higher ratios when the impacted wetlands were especially sensitive, such as when they provided habitat for threatened or endangered species. States that regulate wetlands in addition to federal Section 404 requirements sometimes impose higher mitigation ratios than USACE.

More recently, USACE and states that regulate wetland impacts have focused on no net loss of wetland functions rather than wetland acreage. Under a functional no net loss approach, the required mitigation area may be more or less than one acre per acre lost. The establishment of higher than 1:1 mitigation ratios for forested wetlands or wetlands of special value was an early attempt to factor functionality into the wetland mitigation process. The ability to establish wetland mitigation requirements on the basis of function rather than area requires the availability of methods to assess the functionality of wetlands, both the impacted and the proposed mitigation wetlands. As noted in Chapter 4 of this book, several functional assessment procedures are now available. Different USACE districts (and even different regulatory personnel in the same district) and state regulatory agencies have differing preferences regarding the choice of functional assessment methods. Some states have established functional assessment methods tailored to wetland mitigation activities in that state, for example, the uniform mitigation assessment method (UMAM) now used for most wetland mitigation in Florida.

An example of the mitigation rule, the increased focus on functional assessment, and the increased focus on watershed-focused wetland mitigation is the multilocation wetland mitigation proposal developed for the proposed Levy nuclear reactors in Ingliss, Florida. Although developed subsequent to the mitigation rule, the utility developing Levy, Progress Energy Florida (PEF), is proposing to develop permittee-sponsored mitigation that will essentially form its own mitigation bank.[32] The utility quantified its wetland impacts in terms of UMAM units, reflecting losses of wetland ecological and hydrological function rather than just acreage. The utility then developed a wetland mitigation plan calling for enhancing wetlands on the site as well as four other properties encompassing public and private land in each of five watersheds containing wetlands affected by building and operating the power plant and its off-site pipelines and transmission lines. The plan ensures that the resulting enhancement "lifts" the regional wetland function by at least the same number of UMAM credits as those lost. The Levy plan is one of the first large-scale permittee-sponsored wetland mitigation designs developed subsequent to the mitigation rule. It may therefore serve as a model for postmitigation rule design of permittee-sponsored wetland mitigation.

6.4.3 Endangered Species Act Mitigation

Mitigation for impacts to threatened and endangered species under the Endangered Species Act has many parallels to wetland mitigation under Section 404 of the Clean Water Act. Furthermore, they are not completely mutually exclusive; after all, endangered species as diverse as wood storks (*Mycteria americana*), whooping cranes (*Grus americana*), and ivory-billed woodpeckers (*Campephilus principalis*) all depend on very specific wetland habitats.

6.4.3.1 Avoidance and Minimization

As is true for wetland mitigation, the simplest and most effective mitigation measure can often be just to avoid impacts to endangered species habitats in the first place. As long as the avoidance is complete, this form of mitigation is almost 100% effective and risk free. It is not, however, always practicable. It might be possible to readily route a highway or a transmission line around a nesting site for a listed bird or a clump of a listed plant in some rural areas, but it may be considerably harder if the only available right-of-way crosses urban land and changes might require condemning private residences or businesses. Likewise, it might be possible to direct ground disturbance away from a nesting roost on the edge of a power plant development site but not if the roost lies in the one place possessing the requisite geological or meteorological conditions for the power block or stack. Environmental purists commonly push for avoiding any work in any area containing listed species or critical habitats without understanding the practical limitations faced by project developers. Although avoidance may appear to be almost cost free compared to more complex compensatory mitigation measures, the costs of developing a project around certain places on a site can be substantial. Furthermore, there are "opportunity costs" associated with any decision not to use a given area of real estate for purposes other than preservation of natural habitat.

One may recognize three general approaches to avoidance: spatial avoidance, temporal avoidance, and conditional avoidance. Although I have devised these terms, they present a useful pattern for categorizing possible ways to achieve avoidance of impacts to threatened and endangered species. Each has its advantages under certain circumstances. A well-developed mitigation strategy for a proposed action considers and incorporates elements of all.

6.4.3.1.1 Spatial Avoidance

The most intuitive approach to avoidance is simply to locate proposed disturbances away from habitat containing or potentially benefiting listed species. The discussion in the previous paragraph pertains to spatial avoidance. Ideally, spatial avoidance should begin at a high or regional level and progress iteratively to a localized level. At a high level, avoiding impacts to threatened and endangered species can be part of a comprehensive siting process striving to minimize overall potential environmental impacts. Several organizations have developed standardized siting protocols for specific types of projects. For example, the Electric Power Research Institute (EPRI) has published methodologies for siting new nuclear power plants[33] and transmission lines.[34] The Nuclear Regulatory Commission has published guidance promoting use of that methodology and for integrating into the methodology consideration of ecological resources such as threatened and endangered species.[35]

A common pattern to these methodologies is to begin by defining a large study area where the project could conceivably be situated and then identifying specific opportunities and constraints within that study area (not all methodologies use the terms opportunity and constraint, but most utilize equivalent concepts). Opportunities are features or areas that appear to be noticeably conducive to siting a proposed project while minimizing environmental impacts. Existing industrial parks with easy water access might constitute an opportunity for a proposed port project, while an abandoned railroad right-of-way might constitute an opportunity for a proposed transmission line. Constraints are locations of environmentally sensitive resources, such habitats used by threatened or endangered species. Examples of other constraints include wetlands, floodplains, and historical or archeological sites.

Once a map showing opportunities and constraints is produced, it may be used to identify candidate sites for polygonal development such as power plants or reservoirs or routes for linear development such as highways or transmission lines. Users of the methodologies typically assign numerical weights to each of the opportunities and constraints. Known locations of threatened or endangered species typically receive a relatively high weighting relative to other constraints, although some constraints, such as locations of private residences or certain high-visibility historical sites, are sometimes weighted higher. Because assigning the weighting factors is highly subjective, project proponents commonly employ multidisciplinary teams of specialists who contribute to the weighting process using what is sometimes termed a "Delphi" procedure. The teams typically include at least one biologist and often multiple biologists to provide expertise regarding threatened and endangered species. Sometimes, those biologists seek data from the Services regarding threatened and endangered species using the early "informal consultation" procedures discussed in Chapter 5. But, rarely do the Services participate directly in the siting process at this stage.

Spatial avoidance can be a simple and effective mitigation strategy if practicable. Projects conducted on large government properties and in many rural areas can often be readily directed away from ecologically sensitive habitats. The same is also true in areas containing abandoned or lightly used existing facilities; opportunities often exist to site new facilities within the existing footprints of former facilities rather than in undeveloped natural areas. But, not all projects may be planned with the expectation of such flexibility. Some proposed projects face constraints related to land availability and geotechnical and economic limitations that culminate in a need to site new facilities in sensitive natural habitats. As effective as spatial avoidance may be, it is simply one of the avoidance tools available to the environmental planner.

6.4.3.1.2 *Temporal Avoidance*

Temporal avoidance is refraining from conducting potentially disturbing activities until it is certain that threatened or endangered species are not present in the action area. Temporal avoidance can be a particularly effective way of avoiding impacts to migratory birds, which are known to be present or absent in given areas according to a relatively predictable schedule. For example, the U.S. Forest Service commonly avoids disturbing stands of jack pine (*Pinus banksiana*) in Michigan when the endangered Kirtland's warbler is known to return from the south to breed. The FWS encourages agencies performing actions in parts of the northeastern and midwestern United States to conduct actions disturbing deciduous forest in the winter, when the endangered Indiana bat is known to hibernate in caves rather than to breed and forage in forests. The FWS also encourages action agencies to avoid work within specified distances of bald eagle[36] nesting sites during known breeding and nesting seasons.[37]

Although temporal avoidance can be effective, it is not always foolproof. Many migratory species can display site fidelity when breeding, returning to the same nesting sites in successive years. Many bald eagle nests remain active for many years, with nesting pairs successively enlarging the nest each year. Even if ground disturbance is timed to avoid direct injury to nesting birds and young, the loss of an established nest can disrupt the behavior of birds that return the next year. Individuals of territorial species that lose their territories to land development may be forced to compete with other individuals for new territories. Furthermore, migration patterns are not always completely predictable. There is always at least a small risk that individuals might return to an area or nest outside the predicted window. Changes in climate, whatever the cause, have noticeably altered many seasonal patterns displayed by biota and introduce additional uncertainty into attempting to time actions based on those patterns.

Temporal avoidance is of course not generally effective for most plants and other immobile species such as fungi, coral, or benthic organisms. However, even wildlife species whose adults are mobile during the growing season may experience losses if nesting or hibernation sites are disturbed.

6.4.3.1.3 *Conditional Avoidance*

Conditional avoidance is commonly used for projects affecting nesting habitat for listed migratory birds. If a bird is known to nest during specific seasonal windows, activities damaging their nesting habitat can be conducted outside those seasons to obviate potential physical injury to the reproductive process, eggs, or nestlings. A common mitigation measure for the bald eagle is timing activities for outside known nesting seasons.[38] The specific season of avoidance depends on geography; the breeding season begins and ends progressively later with increasing latitude. Working outside the season does not, however, eliminate all impact. If the work destroys the nesting habitat during the winter season, physical injury is avoided, but migrants returning to nest find their nesting habitat gone and must seek new suitable nesting habitat. Many migratory species, including bald eagles, are territorial at their nesting sites and return to the same sites in subsequent years. Displaced individuals may not find optimal replacement sites to nest, or they may displace other individuals from established territories. Some species, including bald eagles, also add to their established nests in subsequent years; losing an established nest means having to begin building a new nest. This type of avoidance can be even more effective for impacts near, rather than at, a nest. Noise close to a nest may disturb nesting activities even if the nest itself is not disturbed; timing the noise for when the birds are not present may be completely effective in avoiding impact.

6.4.3.1.4 *Other Avoidance Measures*

Although spatial, temporal, and conditional avoidance can be used as broad avoidance categories, other types of avoidance can also be identified. One particularly useful form of avoidance is to design development to minimize encroachment into large areas of forest or other natural habitat and to preserve unbroken strips of natural habitat that function as "travel corridors" for wildlife. This approach for avoidance is particularly well suited to landscape-level regional planning, as occurs when preparing comprehensive land use plans for counties or portions of counties or laying out large-scale mixed-use developments, sometimes termed planned unit developments or PUDs.

One of the most effective approaches to preserving travel corridors is illustrated by Montgomery County, Maryland, a densely developed suburb of Washington, D.C. Even in the most urbanized parts of the county, a series of stream valley parks preserve broad swaths of forest cover encompassing floodplains and side slopes surrounding the stream channels.[39] The unbroken forest cover not only facilitates movement of white-tail deer (*Odocoileus virginianus*) over the otherwise-urbanized landscape but also provides large blocks of forest cover conducive to forest-interior neotropical migratory birds such as scarlet tanager (*Piranga olivacea*) and many wood warbler species. Around the 1990s, black bears (*Ursus americanus*) were observed in forested parklands along Seneca Creek in the Montgomery County suburbs of Gaithersburg and Germantown, more than 50 miles east of the forested ridges where they are more commonly spotted. Many recent PUDs in Montgomery County, such as the widely acclaimed Kentlands community in Gaithersburg, have further striven to incorporate swaths of preserved forest cover within new blocks of high-density residential and commercial development. Even some of the older nineteenth century urban parks, such as Rock Creek Park in Washington, D.C., and Fairmont Park in Philadelphia, serve to provide forested travel corridors for wildlife, even if their designers were not aware of the ecological values of such features.

6.4.3.2 Compensatory Mitigation

Although the concepts of avoidance and compensatory mitigation appear at first blush to be sharply delineable, they are in fact quite gray and gradual.

6.4.3.2.1 Creation, Restoration, and Enhancement of Specialized Habitat

Just as wetlands can be created, restored, or enhanced, so can other types of habitats. This includes specialized habitats, whether wetland or upland, required for endangered or threatened specialists. I was involved in the development of a plan to restore and enhance a specialized type of chaparral habitat on Vandenberg Air Force Base in southern California to offset losses of the habitat needed to refurbish and expand on-base military family housing.[40] The approach was much like a wetland mitigation plan, except that the target habitat was a regionally endemic type of chaparral, Burton Mesa chaparral, which occurs only on or in close proximity to Vandenberg Air Force Base. Burton Mesa chaparral provides specialized habitat for several plant species listed under the Endangered Species Act or by the state of California. The plan called for planting native trees, shrubs, and herbs in areas of former Burton Mesa chaparral that had been eliminated or degraded by grazing and establishment of invasive annual grasses.

6.4.3.2.2 Relocation

Attempting to relocate individuals of threatened or endangered species to suitable receiving habitat outside the action area seems like a logical approach to mitigation, but there are several challenges. Capturing, moving, and releasing individuals into a new setting inevitably imposes substantial stress on the relocated individuals, even if the receiving habitat is not already at carrying capacity for the subject species (the concept of carrying capacity is discussed in Chapter 2 of this book). The relocated individuals are artificially confined for at least a short period and then released to new habitat where they have no established patterns of territoriality. There is uncertainty regarding whether the relocated individuals can actually adapt to and thrive in the new habitat. Still, relocation offers at least a chance of survival in intact natural habitat instead of almost certain death on a development site. Relocation should in most circumstances be viewed as a last-resort desperation measure, a giant heave of the football into the end zone at the end of a game or a full-court hook shot at the end of a basketball game.

As of May 2010, over 900 eastern box turtles (*Terrapene carolina carolina*) had been relocated from the right-of-way of a controversial new freeway project, the Intercounty Connector, in the Maryland suburbs of Washington, D.C., to suitable habitat elsewhere in rural Maryland.[41] The eastern box turtle is not federally or state listed in Maryland, but the Maryland Department of

Transportation has agreed to the relocation effort as mitigation as part of the NEPA process for the freeway. The relocated turtles are being monitored for behavior and survival. Unfortunately, despite the high level of care in the relocation effort, many of the relocated turtles were reported in 2012 to be succumbing to a microbial disease that has ravaged populations of several reptiles in the Mid-Atlantic.[42] As the disease is also affecting individuals not impacted by the freeway project and not involved in the relocation effort, it is unclear whether the stress of the relocation is playing a role in the observed mortality.

6.4.3.2.3 Preservation

The act of setting aside natural habitat for listed species as mitigation for the destruction of other habitat for those species is controversial. Some environmental purists argue that such preservation is not mitigation at all; habitat is lost at the development site, and no offsetting habitat is added anywhere else. More pragmatic environmentalists recognize that unregulated losses of habitat happen almost continually, and that setting aside habitat protected from everyday development is both sensible and effective, indeed in some settings the only way that intact natural habitat occupying expensive land can be saved from development.

The concept of placating environmental interests by setting aside tracts of preserved land in exchange for goodwill pre-dates the Endangered Species Act. One of the most treasured natural areas in southern Florida is the Corkscrew Swamp Sanctuary, owned and managed by the National Audubon Society. Today comprising over 13,000 acres of swampland and pine flatwoods, Corkscrew began in the 1950s with a grant of about 3,000 acres of virgin bald cypress swamp from a logging company to the society. At the time, logging companies had constructed a network of logging trams, or railroads, into the vast tracts of virgin cypress swamps and were rapidly logging out the few remaining tracts of century-old cypress trees. Audubon could not and did not stop the inevitable logging out of the vast swamps, but the logging interests' goodwill gesture today remains an island of what was formerly commonplace over the southern Florida landscape. Corkscrew today provides habitat for a number of listed species, including serving as the premier nesting site for the endangered wood stork. Anyone who doubts the potential mitigative value of preservation should visit Corkscrew.

The Everglades Mitigation Bank owned and operated by Florida Power and Light on property inland from the Turkey Point power plant complex south of Miami is an excellent example of land preservation serving an endangered species, in this case the endangered American crocodile. The bank occupies approximately 13,249 acres encompassing several ecosystems typical of southern Florida that support 46 protected (federal or state) species of plant and animal life, including the American crocodile. The land had originally been purchased in 1970 for power plant development later found to be unnecessary. It forms a seamless corridor of preserved and managed natural habitat between Everglades National Park to the west and Biscayne National Park to the east. It currently offers freshwater herbaceous and forested wetland as well as coastal wetland credits for sale to developers in an accepted service area comprising most of southern Florida.[43] The range of the American crocodile in the United States is very limited, consisting only of coastal areas in extreme southern Florida and the Florida Keys. Hundreds of crocodiles today flourish on the Turkey Point property, including the Everglades Mitigation Bank.

Preservation of natural habitat is especially common and hence controversial in Florida. One reason is of course the availability of substantial areas of natural habitat in a state that retained until the mid-twentieth century broad areas of wilderness largely unaltered by centuries of agriculture and logging. As areas along the southwestern Florida coast were becoming urbanized, opportunities remained to protect large blocks of true wilderness in the nearby interior. Florida also presented a key conundrum sometimes overlooked by the conservationists: Almost everything there is ecologically valuable. Wetlands are more prevalent than nonwetlands (uplands) in many landscapes, and even the driest of interior landscapes are pockmarked seemingly everywhere by isolated depressions and anastomizing strands of linear wetlands. Wetland creation in such landscapes is illogical—there just are not that many uplands to go around. Furthermore, many of the uplands are themselves ecologically valuable and rare. Many federally listed species such as the eastern indigo snake and state-listed species such as the gopher tortoise in fact rely on the availability of these specialized uplands.[44,45] One could argue that preservation of these uplands is more ecologically valuable than preservation of some of the adjoining wetlands.

6.4.3.2.4 Institutional Controls

The following discussion uses the term *institutional controls* to cover a number of diverse mitigation measures that seek to alter human behavior to avoid or minimize effects on threatened or endangered species. Road signs warning drivers of the presence of endangered American crocodiles line the roads inside the Turkey Point Nuclear Power Plant site in Homestead, Florida. The incidental take permit for nuclear operations at the Turkey Point allows for the death of one American crocodile every five years in the form of accidental direct mortality.[46] Most of the individuals lost have been killed by inattentive drivers using roads traversing crocodile habitat on the property. Detroit Edison, which is proposing to build and operate a third nuclear reactor on its Fermi Nuclear Power Plant site in Monroe, Michigan, has proposed to post warning signage and establish speed limits on roadways servicing the new plant to reduce the potential for inadvertent vehicular collisions with the state-endangered eastern fox snake.[47]

Anyone driving I-75 or "alligator alley" across the Everglades and Big Cypress between Miami and Naples has witnessed one approach to engineering controls to protect an endangered species, the Florida panther. Both sides of the highway where it crosses panther habitat are lined by 10-foot tall chain-link fences. Multiple underpasses were built under the highway to allow panthers to pass safely under the highway. The fences help keep panthers from walking out on the highway and direct them to the underpasses.[48]

6.4.3.2.5 *Payments-in-Lieu*

Payments-in-lieu have had a roller coaster of opinion among conservationists. The concept is quite simple: The developer quantifies impact using some metric, such as acres of habitat, functional units using methods such as Florida's UMAM, or numbers of individuals taken; then, the developer pays some proportional payment to a state or private conservation organization to do with as it sees fit. Ideally, the receiving party would use the money to perform the appropriate mitigation that might otherwise be expected from the developer, usually the creation, restoration, or enhancement of a targeted habitat. Of course, the money could be apportioned to administrative expenses, public relations, or some other less-tangible benefit; the ultimate success of the process is clearly reliant on the integrity of the receiving party. Developers often like payments-in-lieu because they are one-time donations that can be easily planned and incorporated into project balance sheets with no long-term commitments to monitoring and without the long-term uncertainty of ensuring that the mitigation project achieves objectives set by regulatory agencies. Critics of course view payments-in-lieu as essentially a pay-for-impacts approach and worry that recipients of the payments may not properly steward the payments into the appropriate mitigation objectives.

For most of the 1990s and 2000s, payments-in-lieu were generally viewed by conservationists and environmental regulatory agencies as a last resort, to be encouraged only when other more direct mitigation measures were impossible or impracticable. Payments-in-lieu had certain elements in common with mitigation banking, which became ever more favored over that period. Specifically, mitigation banking also involved transferring the responsibility for performing the mitigation from the developer to outside interests with specialized expertise. But, the regulatory agencies could maintain tight control over the banks, approving bank credits for sale only when the banking projects demonstrated success in attaining desired ecological benefits. Payments-in-lieu were quite different. Conservation interests received money, not habitat. Money is too fungible; nothing was in place to ensure that the money was properly shepherded into establishing successful habitat. And, if the money were squandered on poor habitat improvement projects, or worse frittered into "general administrative funds," there was no way to go back to the developer. The impacts were final, and the money was squandered.

When the mitigation rule was ultimately issued in 2008, many conservationists were therefore surprised when the rule actually favored payments-in-lieu over traditional permittee-sponsored mitigation, even on-site, in-kind mitigation. The rule, as expected, favors use of mitigation banks, but if banks are not practicable, it encourages consideration of payments-in-lieu before considering permittee-sponsored mitigation. Most expected that the rule would favor banks, but they did not expect it to revitalize the generally discredited payment-in-lieu approach. The rule is too new at this point, especially considering the downturn in new construction since the 2008 recession, to know if payments-in-lieu will become an effective alternative to banks or a loophole that weakens mitigation objectives.

6.4.4 The Future of Mitigation

Whether for endangered species, wetlands, or other ecological resources, the concept of mitigation, at least the concept of compensatory mitigation, divides environmental consultants into two camps. One camp thinks that the only effective mitigation is avoidance, that the original resources are so complicated and so poorly understood that the only way to conserve the benefits of those resources is to preserve the original resource itself. The other camp advocates that the principles of science and engineering and our inherent innovative capacity can be applied to establish reasonably good facsimiles of natural habitats. The distribution and patterns of plants, soils, and hydrology in the targeted habitats can be studied prior to their destruction and copied onto blueprints that can be used to replicate the habitats closely, at least over time. The targeted species would not be able to tell the difference.

The jury is of course still out on which camp is correct, and most likely both are. I have designed tidal wetland creation projects at Dahlgren, Virginia, that at least appear to remarkably simulate the appearance of natural wetlands. Planted sprigs of regionally indigenous tidal marsh species found in adjoining intact tidal marshes such as *Spartina alterniflora* and *Spartina cynosuroides* are thriving in the same positions on the intertidal gradient as in the natural marshes used as models in the design. For more complex natural systems such as nontidal forested wetlands on Dahlgren, the results have been encouraging but less certain. The biggest uncertainty is, of course, the trees; even under exactly the right conditions, the small saplings will take more than 30 years to reach mature heights. Less obvious is whether the cluster of mature trees, after the 30 years, truly have the spatial and functional characteristics of a natural forest dominated by those same tree species, or simply will be a cluster of mature trees lacking those characteristics. After all, a cluster of pine or oak trees in the midst of a lawn is not really a fully functioning forest unless it establishes not only the understory and groundcover but also the complex fob web present in natural forests.

Many wildlife experts consider natural habitat to be irreplaceable. Others marvel at how their knowledge can be applied in an engineering capacity to establish modern marvels such as wildlife underpasses or wetland creation to conserve wildlife and allow for modern progress at the same time. The fact is that the latter are more on; progress really is inevitable, we live in an increasingly crowded world and suburban development—even the so-called dreaded suburban sprawl—will occur. The question is how we can reconcile conservation of endangered species and other wildlife into the increasingly urbanized world that we are powerless to prevent. The sight of hotels set more than a quarter mile from the beach, separated by mangroves, and trams carrying beachgoers across the mangroves in electric trams—no carbon emissions—really should be a delight to conservationists. People will always want to go to the beach, but in this way they will also get to appreciate the mangroves as they go. And, the wildlife will continue to benefit from the mangroves as they always have.

For environmental consultants, mitigation is perhaps the most exciting opportunity for creativity in meeting technical challenges. Whether in the form of mitigation banks, projects sponsored by parties receiving payments-in-lieu, or permittee-sponsored mitigation projects, mitigation is perhaps the most demonstrable venue for actually protecting and improving the environment while facilitating the growth on which our quality of life depends. Describing how proposed projects can adversely affect threatened and endangered species will contribute to recovery of those species only if it leads to informed decision making among practicable alternatives, avoidance of easily avoided impacts, and channeling our collective intellectual faculties into developing innovative compensatory mitigation solutions.

Notes

1. 50 C.F.R. 402.02.
2. The regulations require consulting action agencies to prepare and submit a biological opinion only for "major construction actions." Other actions are still within the scope of Section 7, but action agencies may provide requisite project description data on those actions using a venue other than a biological assessment, such as a cover letter with attachment.
3. 50 C.F.R. 402.02.
4. U.S. Fish and Wildlife Service and National Marine Fisheries Service. 1996. *Habitat Conservation Planning and Incidental Take Permit Processing Handbook.* U.S. Department of the Interior, Fish and Wildlife Service, and U.S. Department of Commerce, National Oceanographic and Atmospheric Administration, National Marine Fisheries Service, Washington, DC, November 4, p. 4-13.
5. 50 C.F.R. 402.02.
6. 40 C.F.R. 1502.14.
7. NEPA's 40 Most Asked Question, Response to Question 2a.
8. 40 C.F.R. 230.10(a).
9. 33 C.F.R. 328.
10. 40 C.F.R. 230.10(a)(2) and (3).
11. 16 U.S.C. 1536(b)(4).

12. U.S. Fish and Wildlife Service, *Habitat Conservation*, pp. 4-13.
13. Ibid., pp. 4-47.
14. Ibid., p. xvii.
15. Ibid., pp. 4-59.
16. 50 C.F.R. 17.22(b)(1).
17. U.S. Fish and Wildlife Service, *Habitat Conservation Planning*.
18. U.S. Fish and Wildlife Service, *Habitat Conservation*, preface.
19. 40 C.F.R. 1502.20.
20. Regulations for the use of general permits under Section 404 are contained in 33 C.F.R. 330, and the list of current general permits is contained in Appendix A to 33 C.F.R. 330.
21. The concept of general permits under Section 404 bears some similarities to the concept of categorical exclusions under NEPA. By extension, since regional habitat conservation plans bear similarities to Section 404 general permits, one may conclude that the former also bears similarities to categorical exclusions.
22. 63 FR 8859-8873, February 23, 1998.
23. 64 FR 32717, June 17, 1999.
24. 71 FR 15213-15215, March 27, 2006.
25. U.S. Fish and Wildlife Service, *Habitat Conservation*, Chapter 1.
26. 40 C.F.R. 1508.20.
27. 10 C.F.R. 1021.331.
28. NEPA's 40 Most Asked Question, Response to Question 40.
29. NEPA Task Force. 2003. *The NEPA Task Force Report to the Council on Environmental Quality: Modernizing NEPA Implementation*, September, Chapter 6. Washington, DC: Author.
30. Environmental reports commonly refer to the ground surface substrate of terrestrial uplands and wetlands with persistent emergent vegetation as "soils," while referring to the substrate of water bodies and wetlands with only submerged vegetation or nonpersistent vegetation as "sediments."
31. *Federal Register*, 73(70), 19594–19705 (2008). Compensatory Mitigation for Losses of Aquatic Resources; Final Rule. April 10.
32. U.S. Nuclear Regulatory Commission. 2011. *Draft Environmental Impact Statement for Proposed Levy Nuclear Power Plant*, Office of New Reactors, Rockville, Maryland.
33. Electric Power Research Institute (EPRI). 2002. *Siting Guide: Site Selection and Evaluation Criteria for an Early Site Permit Application*, EPRI, Palo Alto, CA.
34. Electric Power Research Institute (EPRI) and Georgia Transmission Corporation (GTC). 2006. *EPRI-GTC Overhead Electric Transmission Line Siting Methodology*, EPRI, Palo Alto, CA, and Georgia Transmission Corporation, Tucker, GA.
35. U.S. Nuclear Regulatory Commission. 1998. *Regulatory Guide 4.7—General Site Suitability Criteria for Nuclear Power Stations*, Rev. 2, April. Appendix B. Rockville, MD: Author.
36. The bald eagle is now delisted from the Endangered Species Act but is still protected by FWS under the Bald and Golden Eagle Protection Act.
37. U.S. Fish and Wildlife Service. 2007. *National Bald Eagle Management Guidelines*, U.S. Fish and Wildlife Service, Washington, DC, May 2007, 23 pp.
38. Ibid.

39. Maryland Greenways Commission. 2000. Maryland Atlas of Greenways, Water Trails, and Green Infrastructure. Available at http://www.dnr.state.md.us/greenways/counties/montgomery.html. Accessed March 17, 2012.

40. U.S. Air Force. 1996. *Final Burton Mesa Chaparral Mitigation Plan for the Replacement of Military Family Housing Phases 3-14*, March , U.S. Air Force, Vandenberg Air Force Base, Lompoc, CA.

41. Hogan, Terri. 2010. 900 Turtles Relocated from Path of ICC Project; Part of Initiative to Protect Native Maryland Species. Gazette.com. Maryland Community Newspapers Online. May 12. Available at http://ww2.gazette.net/stories/05122010/olnenew223327_32554.php. Accessed March 17, 2012.

42. Shaver, Katherine. 2012. Deadly Ranavirus Hits Box Turtles, Tadpoles in Montgomery County, Maryland. Wildlife Response.org, February 12. Available at http://www.wildliferesponse.org/index.php/Latest/deadly-ranavirus-hits-box-turtles-tadpoles-in-montgomery-county-maryland.html. Accessed March 18, 2012.

43. Florida Power and Light. Everglades Mitigation Bank. Available at http://www.fpl.com/environment/emb/availability.shtml. Accessed February 20, 2012.

44. U.S. Fish and Wildlife Service. 1982. *Recovery Plan for Eastern Indigo Snake*, U.S. Fish and Wildlife Service, Atlanta, GA, 23 pp.

45. Gophertortoise.org. Gopher Tortoise Facts. Available at http://www.gopher-tortoise.org/tortoise/facts.htm. Accessed February 20, 2012.

46. U.S. Nuclear Regulatory Commission (NRC). 2006. Letter dated June 29 from Frank Gillespie, Director, Division of License Renewal, NRC Office of Nuclear Reactor Regulation, Rockville, MD, to Paul Souza, Acting Field Supervisor, South Florida Ecological Services Office, U.S. Fish and Wildlife Service, Vero Beach, FL. Available at http://pbadupws.nrc.gov/docs/ML0618/ML061800100.pdf. Accessed March 17, 2012.

47. U.S. Nuclear Regulatory Commission. 2011. *Draft Environmental Impact Statement for Proposed Fermi 3 Nuclear Power Plant*, Office of New Reactors, Rockville, MD.

48. Florida Wildlife Corridor. 2012. Day 16: Florida Panther National Wildlife Refuge. Available at http://www.floridawildlifecorridor.org/blog/. Accessed February 20, 2012.

7

The Endangered Species Act and the States

7.1 Introduction

Anyone seeking to understand U.S. environmental laws and regulations (or other laws and regulations) would do well to consider government in the United States as being tiered. The upper tier is the federal government. Federal laws are passed by the Congress, signed by the president, and administered by federal agencies. The second tier is the states. State laws, which are only in force within the corresponding state, are passed by state legislatures and passed by governors. The relationship between the states and the federal government is intricate; states are empowered by the Constitution with high levels of autonomy, and the Tenth Amendment in particular reserves powers not specifically granted by the Constitution to the federal government to the states. The third tier is at the local (municipal) level: counties, towns, townships, cities, and so on. These local jurisdictions also enact laws covering activities within their boundaries. States differ widely in how local governments are established and empowered.

The remainder of this chapter is divided into only two sections. Section 7.2 is a broad overview of state-level regulation directed at protecting endangered (and threatened) species in the United States. Section 7.3 is a more detailed summary of state-level regulations for endangered (and threatened) species for three states that are illustrative of the widely varying differences in state approaches. Note that, among those states that have established state-level statutes protecting rare species, some states designate and regulate both endangered and threatened species, while others designate and regulate only endangered species. The regulatory terminology varies greatly among states and does not always parallel the federal terminology.

7.2 Overview of State Endangered Species Regulation

As a federal law, the Endangered Species Act is in force throughout the United States, in all states, territories, and the District of Columbia. The reach of the Endangered Species Act is independent of whether a state has state-level endangered species regulations. Many species designated as endangered or threatened under the Endangered Species Act bear those designations only in specified geographic ranges, but these geographic limits to federal statuses are based on ecological factors related to the geographical distribution of the species and are independent of what, if any, protections are extended to those species by the affected states. Any state-level protections are *in addition to* (and not instead of) applicable federal protections. States establish state-level protections at their discretion, but states do not have the discretion to reduce, ignore, or override protections under the Endangered Species Act. This chapter focuses on protections extended by the individual states to endangered species, independent of the Endangered Species Act.

Many federal environmental laws—but not the Endangered Species Act—include provisions for allowing states to assume responsibilities for administration of the laws (or portions of the laws) from the federal government, as long as the state agrees to meet minimum standards of protection consistent with the objectives of the law. The federal government is said to "delegate" these laws (or portions of laws) to the states, which administer the delegated laws within their respective boundaries. The environmental law best known for delegation to the states is the Clean Water Act. All states but a few (e.g., Massachusetts) have asked for and received delegation of Section 402 of the Clean Water Act, better known as the National Pollutant Discharge Elimination System (NPDES). States taking on NPDES, which establishes permits for liquid discharges to waters of the United States, must agree to enforce standards on chemical concentration and other physical characteristics (e.g., temperature and pH) that are at least as stringent as those set by the U.S. Environmental Protection Agency (EPA). Many states have established stricter standards than those set by the EPA; that is acceptable, but setting standards weaker than EPA's would cause a state to lose delegation. Two states, Michigan and New Jersey, have also been delegated Section 404 of the Clean Water Act, which regulates solid discharges (fill) into wetlands and waters of the United States. Delegation of environmental laws not only is consistent with the objectives of maximal state autonomy (federalism) established by the Constitution but also allows for administration of the law by agencies more familiar with regional environmental conditions and more directly answerable to the local populace.

However, there are no provisions for delegation of the Endangered Species Act to states. Many of the arguments supporting state delegation of the Clean Water Act would seem to apply equally well to delegation of the Endangered Species Act, especially when one considers that the Endangered Species Act has even more potential to adversely affect the property rights of local landowners than the Clean Water Act. However, the fact that so many threatened and endangered species are migratory, moving over large geographic areas over the course of their life cycle, argues for more centralized administration. Furthermore, administration of the Endangered Species Act requires the availability of a large staff of subject matter experts who can be more practicably assembled at a federal agency such as the Fish and Wildlife Service (FWS) and National Marine Fisheries Service (NMFS) than at all but the largest of states. Whether current popular demand for increased regulatory flexibility results in Congress passing legislation delegating all or parts of the Endangered Species Act to the states remains to be seen. So far, Congress has not established the regulatory framework for delegating the act to the states. Even if Congress were to do so, few states are likely to welcome the challenge and expense of effectively administering the objectives of the Endangered Species Act.

Nevertheless, most states have some state-level legislation addressing the protection of species. Most states have developed state lists of threatened or endangered species. State lists are not lists of federally listed species that occur in the state. They are not the same as the lists of federal threatened and endangered species occurring in each state published by FWS. Many states automatically consider each federally listed species occurring in the state to be state listed, but most also establish state-listed species that are not federally listed. Many nonfederally listed species included on state lists are in jeopardy of extirpation from the state (state endangered) or are at risk of becoming in jeopardy of extirpation from the state (state threatened). Others are species for which the states have concluded that protections are warranted despite the fact that the federal review process has not resulted in the extension of federal protections.

The exact definitions of the words *endangered* and *threatened* in state legislation vary somewhat among states, as do the criteria for designation as such. In addition, some states (e.g., California) designate state candidate species or species conceptually similar to the concept of candidate species. Some states (e.g., Florida) designate endangered and threatened species as well as a pool of other rare species designated as species of state concern (or some similar version indicating state concern). Many state statutes establish criteria similar to those of the Endangered Species Act for inclusion on their lists, except that they address possible extirpation from the state in addition to possible global extinction.

Note the use in the previous text of the word *extirpation*. Many states speak of species in danger of "extinction" from the state, but that language is technically incorrect. Extinction means extinct everywhere; elimination of a species from some spatially defined subarea (such as a state) but without extinction everywhere is termed extirpation. Political boundaries, including those of states, have no ecological meaning. Elimination of a species from a state may have emotional significance to residents of that state, but if healthy populations of the species remain elsewhere, the overall impact on the species is open to debate.

There is a common misconception regarding species designated as threatened or endangered by a state. The criteria used by states are generally based on rarity *in the state*, not just overall rarity. Not all species designated as threatened or endangered by one or more states are actually rare or facing possible extinction. They need only be rare or facing possible extirpation from the subject state. An example is the American lotus (*Nelumbo lutea*), an aquatic plant that is very common in shallow waters and wetlands in many southeastern states. It is listed as threatened by Michigan and endangered by New Jersey and Pennsylvania but as a noxious weed by Connecticut.[1] The American lotus is common in freshwater marshes in many southeastern states. Michigan encourages transplantation of American lotus specimens in the way of proposed development, while Connecticut prohibits planting the species. Of course, many species listed by states are rare overall, and many federally listed species are also listed by one or more states. However, not all federally listed species occurring in some states are necessarily listed by that state. Remember that there is an element of politics as well as science in decisions to list species, at both federal and state levels.

Only a few states, specifically Alabama, North Dakota, West Virginia, and Wyoming, have no statutes addressing threatened or endangered species. Some states (e.g., Alabama) designate listed species for the state but do not regulate activities affecting those species. Many states designate listed species and regulate commerce in and intentional takings of those species but have no provisions addressing incidental take or otherwise limiting impacts on those species from development, especially private-sector development. Only a few states (e.g., California and New Jersey) regulate incidental take in a manner at least somewhat resembling the Endangered Species Act. The extent to which individual states establish and expand state-level lists of endangered species and regulate impacts on those species is generally reflective of the political culture of the states but can also reflect the prevalence of rare species in the states and threats to those species. In general, states with the most aggressive state-level regulatory protections for state-listed species tend to be both politically liberal and burdened with the highest population densities. Examples of such states include Massachusetts, New Jersey, and Maryland. Even though portions of Texas have experienced rapid and dense urban development, Texas is generally a libertarian state with a culture exceptionally geared to individual liberty and property rights. Even though portions of California are sparsely populated, California tends toward centralized government and careful land use planning. Nevertheless, no state is openly opposed to protecting its natural heritage, including rare and especially endemic species, and developers and other project proponents in even the most libertarian of states should count on having to take reasonable efforts to plan to minimize impacts on environmental resources and document those efforts, together with reasonable mitigation measures, to appropriate state regulatory authorities.

Despite the variability in how individual states regulate threatened and endangered species, a few generalizations may be made regarding state threatened and endangered species regulatory programs:

1. Species listed as threatened or endangered by states may be common overall, and not all federally listed threatened or endangered species occurring in a state are listed by that state.

2. State endangered species acts tend to be less rigorous and less tightly enforced than the Endangered Species Act. There are exceptions, such as New Jersey and California.

3. State endangered species acts tend to be more weakly funded and less well staffed than the U.S. FWS, sometimes resulting in long waits to receive requested information or reviews. This is true even in some states with tight regulatory requirements.

4. State endangered species regulations tend to focus more on providing guidance and recommendations and less on permitting or enforcement.

5. Most states shy away from placing limitations regarding threatened or endangered species on private development or private property.

6. Many states offer more protections for threatened or endangered species of wildlife than plants.

States vary considerably in how many species they include on state lists. The list of state-listed species for Alaska as of January 2012 totaled only five species. Table 7.1 is a summary of the state-listed species for Florida. The reasons for the differences are both scientific and political. From a scientific perspective, northern settings such as Alaska are generally lower in biodiversity than nondesert settings such as Florida, which are in or close to the tropics. As explained in Section 7.3.2, Florida also possesses several unique natural habitats, found nowhere else in the world and supporting many endemic species, that are unusually sensitive to disturbance and that have experienced aggressive agricultural development followed by explosive urban growth. Alaska, by contrast, still contains very large areas of wilderness that have experienced relatively little overall human disturbance (although localized areas of Alaska have experienced heavy disturbance). From a political perspective, Alaska still considers itself to have frontier qualities, with seemingly unlimited natural areas, and depends on development of natural resources, especially oil and gas, to support its economy. Florida, by contrast, depends heavily on tourism for its economy, and tourism is benefited by attractive and intricate natural areas. Wilderness in Florida is largely confined to a few publicly protected natural areas such as Everglades National Park and Big Cypress National Reserve, with the remainder of its landscape largely agricultural or urban. One unifying theme illustrated by both Alaska and Florida, however, is that the degree to which states establish regulatory protections for threatened or endangered species (beyond minimal compliance with federal regulations under the Endangered Species Act) is driven to a large extent by economic considerations.

TABLE 7.1

Breakout of Fish and Wildlife Listed under Florida Endangered Species Act

Status	Fish	Amphibians	Reptiles	Birds	Mammals	Invertebrates	Total
Federal endangered	3	1	4	9	22	7	46
Federal threatened	1	1	6	4	1	6	19
Federal threatened (S/A)	0	0	1	0	0	0	1
Federal extinct	0	0	0	1	0	0	1
State threatened	3	0	7	5	4	2	21
State special concern	7	4	6	16	6	4	43
Total	14	6	24	35	33	19	131

Source: From Florida Fish and Wildlife Conservation Commission. 2011. *Florida's Endangered and Threatened Species.* October, 10 pp, p4. Tallahassee, FL: Author.

It is important not to confuse regulation of federally listed threatened or endangered species under the federal Endangered Species Act with regulation of state-listed threatened or endangered species. Just because a state has no state-level regulations addressing threatened or endangered species does not mean that the Endangered Species Act does not apply there. Just because state regulations may not be as stringent as the Endangered Species Act does not mean that the federal act is enforced any less strictly there. State lists of threatened or endangered species are developed independently of federal lists. Species on one list are not necessarily included on the other, and if they are on both lists may have differing statuses (e.g., federally threatened and state endangered). Any requirements imposed by a state are in addition to any imposed under the Endangered Species Act.

The key message of this chapter is the following: Regardless of where a project is proposed, project developers and natural resource managers need to be aware not only of the Endangered Species Act but also of state laws protecting threatened and endangered species. The state laws not only can apply as additional protection to many of the same species protected under the Endangered Species Act but also can extend to species not covered under the Endangered Species Act.

7.3 Examples of State Endangered Species Acts

There is not enough room in this book to summarize the endangered species regulations for every state. Instead, the following sections summarize the regulations for three states, providing a representative cross section of state approaches to endangered species regulations:

- Maryland is a relatively progressive state with respect to environmental regulation and is typical of the approaches taken by many northeastern, northern midwestern, and West Coast states.

- Florida is likewise relatively progressive but considerably more pragmatic; its motivations toward environmental regulation are driven more by its tourism-based economy and challenges faced with respect to extremely rapid growth and limited supplies of freshwater. As demonstrated famously in recent elections, Florida is also an especially divided state, balancing conflicting interests of its many urbanized areas in the southern and Atlantic Coast areas with its more rural areas in the northern and Gulf Coast areas.

- Texas places an even higher emphasis on property rights and individual freedom, although it does not fully discount the value of environmental conservation. Texas has still established meaningful state-level protections for endangered species.

The summaries presented are for analytical purposes only. Readers should contact the states or other published sources for up-to-date information on state regulatory requirements.

7.3.1 Maryland

Maryland is a small but ecologically diverse state that encompasses TBD ecoregions (geographical areas with distinct ecological characteristics) extending from sandy beaches, vast tidal marshes, rolling forested uplands, and dry forested ridges. Part of the deciduous forest formation covering most of the eastern United States, Maryland's forest cover consists of distinct belted forest regions proceeding from east to west. According to Lucy Braun,[2] the lowest elevation coastal plain lands adjoining the Atlantic Ocean and Chesapeake Bay are part of the oak-pine forest region, typical of parts of the southeastern United States. The rolling hills and ridges of the midsection of the state are part of the oak-chestnut forest region, typical of piedmont areas in states north and south of Maryland. The highest-elevation Appalachian highlands in the far western part of the state are part of the mixed meso-phytic forest region more typical of highlands to the north.

Aquatic habitats include nearshore sea, estuaries, nontidal streams and rivers, and human-made lakes and ponds but no natural lakes. Extensive wetlands consisting mostly of tidal marshes and blackwater swamps lie to the east of the Chesapeake Bay, while the western shore of the bay is domi-nated in many places by steep bluffs and narrow zones of wetlands. Most wetlands in the middle and western parts of the state are narrow and local-ized, especially in floodplains and along seeps and drainages.

As a densely populated state with two major metropolitan areas and inten-sive agricultural activity, Maryland has a long history of tight environmental regulation. Much of the state's environmental regulatory processes were estab-lished directly or indirectly in response to the public's desire to protect the Chesapeake Bay. The physical and cultural landscape of much of Maryland is dominated by the bay and its associated tidal creeks and inlets, which bisect the state into the nearly flat eastern shore to the east and hilly (and in places mountainous) Maryland mainland to the west. The mainland encompasses the steeply dissected upper portion of the coastal plain immediately west of the bay (sometimes termed the western shore); the hilly and rolling piedmont in the middle section of the state (which includes Baltimore and the Maryland suburbs of Washington, D.C.); the steeply alternating ridges and valleys to the west; and the high-elevation Appalachian highlands in the extreme west.

7.3.1.1 Statutes and Regulations

Maryland protects state-listed threatened and endangered species through the Maryland Nongame and Endangered Species Conservation Act (Maryland Natural Resource Article 10-2A-01-09). The language in the statute generally echoes that of the Endangered Species Act. Following its definition, the statute opens by stating (10-2A-02) the following:

1. It is the policy of the state to conserve species of wildlife for human enjoyment, for scientific purposes, and to ensure their perpetuation as viable components of their ecosystems;

2. Species of wildlife and plants normally occurring within the state that may be found to be threatened or endangered within the state should be accorded the protection necessary to maintain and enhance their numbers; and

3. The state should assist in the protection of species of wildlife and plants that are determined to be "threatened" or "endangered" elsewhere pursuant to the Endangered Species Act of 1973, 87 Stat. 884, by prohibiting the taking, possession, transportation, exportation, processing, sale, offer for sale, or shipment within the state of endangered species and by carefully regulating these activities with regard to the threatened species.[3]

The state act specifically seeks to compliment the federal act. Section 10-2A-04 of the act specifically identifies all endangered or threatened species listed under the federal act as automatically protected under the same status under Maryland's act. However, the act allows for state listing of additional endangered and threatened species based on their population levels within the state. As is the case for species listed by several other states, not all species listed by Maryland are necessarily rare overall or in danger of extinction; many are simply rare or in danger of extirpation from areas within the state boundaries of Maryland. The act also allows for designation of endangered state status for species federally listed as threatened.

The state lists endangered species, threatened species, and species in need of conservation.[4] Endangered and threatened species are defined in a way generally similar to the federal definitions (but from the perspective of Maryland's boundaries). Species in need of conservation are defined by regulation as those "determined by the Secretary to be in need of conservation measures for its continued ability to sustain itself successfully."[5] They may be thought of as generally similar to species federally designated by the FWS as "species of special concern" or the "species of special state concern" identified by many states as a lower tier to endangered or threatened. However, unlike the federal government or most states, Maryland actually regulates impacts to species in need of conservation, although no permits are required for incidental take of those species.[6] Maryland has no process for issuing incidental take permits for endangered species; developers are expected to avoid impacts to those species. They do, however, have a process for issuing incidental take permits for threatened species.[7]

7.3.1.2 Other Related Maryland Statutes

7.3.1.2.1 Maryland Chesapeake Bay Critical Area Act

Perhaps the most unique environmental statute in Maryland is a statute enacted in 1986 to regulate land development in a narrowly defined "critical area" generally extending 1000 feet from the Chesapeake Bay.[8] Specifically, the act defines the critical area as extending 1000 feet landward of the mean high-tide elevation of the Chesapeake Bay and its tidal tributaries, to the head of tide. The act designates a Critical Area Commission to oversee the act but authorizes local jurisdictions (counties and incorporated municipalities) to administer the act and issue approvals. The act does not discourage all land development in the critical area; rather, it divides the critical area into three categories deemed to have scalable needs for protection: intensive development areas (IDAs), limited development areas (LDAs), and resource protection areas (RPAs). IDAs tend to be areas with a history (pre-1986) of urban or industrial development, while RPAs tend to be in rural areas with wetlands, forests, and other natural vegetation. Most threatened and endangered species are of course more prevalent on RPAs than IDAs. LDAs tend to be intermediate in character, often consisting of patches of natural vegetation land adjoined by residential or other lower-density urban development. Identification and mapping of IDAs, LDAs, and RPAs took place at the local (county or city) level, with oversight by the Critical Area Commission.[9]

The Critical Area Act is more of a land use policy than an environmental permitting program. It does not target no encroachment by development, even in RPAs. The act establishes targets of progressively more restricted development for RPAs compared to LDAs and LDAs compared to IDAs. Local jurisdictions incorporate the objectives of the act into their comprehensive land use plans, and most proposed development projects are permitted locally rather than directly through the Critical Area Commission. The act generally seeks low-density development, commonly on the order of one dwelling unit per 20 acres, in RPAs but allows for and even encourages clustered development on smaller lots, leaving broad intervening areas of undeveloped natural habitat.

The act provides sweeping protections to land in immediate proximity to the Chesapeake Bay but does not address land development activities contributing to sedimentation and runoff by way of nontidal tributaries flowing into the bay from distant watershed areas in Maryland and other states. The bay's watershed includes large areas not only of Maryland and Virginia but also of Pennsylvania, West Virginia, New York, and Delaware. Some owners of property on the bay's shoreline have complained that the act places a disproportionate burden on them relative to owners elsewhere in the watershed. Since enactment of the Critical Area Act in 1986, Maryland enacted the Nontidal Wetlands Protection Act in 1991 (discussed separately in the next section), which regulates development in nontidal wetlands and adjoining buffers over the remainder of Maryland. The 1991 act thereby extends many but not all of the critical area protections to similarly sensitive inland wetlands and adjoining buffers in the bay's watershed. Pennsylvania and Virginia have similar state-level protections for nontidal wetlands but not buffers, but West Virginia and Delaware do not regulate nontidal wetlands at the state level. Although most of the tidal shoreline of Maryland is associated with Chesapeake Bay, a smaller area of tidal shoreline adjacent to the ocean estuaries directly connected to the Atlantic Ocean is regulated in a similar way under a complementary act.

One issue of importance under the act is preservation of habitat for forest interior-dwelling birds. The act was ahead of its time in 1986 when it issued guidance on surveying for such birds and preserving their habitat on development sites in the critical area.[10] Since then, the issue of protecting habitat for forest interior birds has become increasingly controversial, as populations of many once-common birds have become markedly reduced.[11] Fragmentation of formerly contiguous swaths of forest cover by housing developments, highways, utility rights-of-way, and other urban amenities has resulted in many small patches of forest cover that are poorly suited to species requiring extensive forest interiors and made it increasingly difficult for such species to move about the landscape. Although few such birds are presently listed as threatened or endangered under the Endangered Species Act or by Maryland, several such species, for example, the cerulean warbler (*Dendroica cerulea*), are presently under evaluation for possible listing. Many of the affected species are neotropical migrants that winter in Central and South America and migrate to the eastern United States in spring and summer to breed. Many of the affected species are once-common species that are highly colorful in their summer (breeding) plumage and highly enjoyed by bird-watchers. Other examples include various other warblers, such as the hooded warbler (*Wilsonia citrina*), prothonatary warbler (*Protonotaria citrea*), yellow-throated warbler (*Dendroica dominica*), and worm-eating warbler (*Helmitheros vermivorum*), and the scarlet tanager (*Piranga olivacea*).

Few of these species have been listed under the Endangered Species Act. The cerulean warbler, identified under the act as a species of special concern, was reviewed by the FWS for listing as threatened, but the FWS concluded that as of 2006 the species did not warrant listing.[12] One noteworthy exception, which never ranged into Maryland, is the ivory-billed woodpecker (*Campephilus principalis*), whose population reductions are thought to have largely resulted from loss and fragmentation of broad, uninterrupted swaths of old-growth swamp forest in the southern United States. The decreasing availability of large uninterrupted tracts of forest or of other natural habitat is a threat to many listed species not only in densely developed states like Maryland but also throughout the United States and the world.

7.3.1.2.2 Maryland Wetlands Acts

Maryland has established separate state-level protections for tidal and other (nontidal) wetlands. Maryland established a Tidal Wetlands Act in 1970 and a nontidal wetlands protection act in 1991. The tidal act protects tidally submerged waters and tidal flats as well as vegetated tidal wetlands. The state limits the tidal act to areas specifically designated as tidal wetlands (or state tidally submerged waters) on state maps, although applicants may petition for updates to the maps. The criteria for inclusion on maps are based on the presence of specific plant species and hydrological criteria but do not follow the three-parameter wetland delineation procedure used at the federal level for Section 404. The tidal act was enacted as part of the well-known wave of environmental protection legislation enacted in the early 1970s. The Maryland act was spurred in part by the highly visible filling of tidal marshes in Ocean City, Maryland, along the Atlantic beaches.

The interrelationships between nontidal wetlands and the estuarine habitats of the Chesapeake Bay and nearshore Atlantic Ocean were not understood in the early 1970s, certainly not well enough understood by the public at that time to override public concerns over property value diminutions resulting from regulatory wetland protections. Unlike the tidal act, the nontidal act is not limited in its scope to wetland features depicted on regulatory maps. Instead, its scope extends to any wetlands meeting the three-parameter federal wetland delineation criteria. Site-specific delineations are necessary, although the same delineation generally suffices for both state and federal wetland protection regulations. Unlike Section 404, whose scope extends not only to actual wetlands but also to other waters of the United States, the Maryland act is targeted specifically to nontidal wetlands. However, other state laws protect stream and river channels and other nontidal open waters of the state. Between these other state laws and the tidal and nontidal wetlands acts, Maryland extends state-level regulatory protections to nearly any wetland or surface water feature covered by federal law. Furthermore, the Maryland wetlands acts are not restricted from regulation if hydrologically isolated from navigable waters, although the state offers some expedited permitting for impacts to isolated wetlands.

Both Maryland wetlands acts protect their respective wetland type in a manner similar to Section 404 of the Clean Water Act. However, the Maryland statutes also extend regulation to nonwetland "buffer" lands landward of actual wetlands. The buffers are typically 25 feet for nontidal wetlands and 100 feet for tidal wetlands but can extend as much as 150 feet depending on local (usually county) restrictions and whether a wetland is specifically designated as "of special state concern." The known presence of federal or state-listed threatened or endangered species is one criterion used in designation of wetlands of special state concern. Protection of wetlands and buffers offers considerably more overall protection to state or federal threatened and endangered species than does protection of wetlands only (as under Section 404). Few threatened or endangered species specifically depend on preservation of tidal marsh habitats, but several such species utilize buffer areas immediately landward of the marshes. Examples include the Puritan tiger beetle (*Cicindela puritana*) and Delmarva fox squirrel (*Sciurus niger cinereus*). Of course, preservation of tidal marshes plays a key role in protecting these shoreline habitats from erosion and sedimentation. Many listed species, such as swamp pink (*Helonias bullata*) and eastern bog turtle (*Glyptemys muhlenbergii*), depend primarily on nontidal wetland habitats, but protection of adjoining buffers contributes substantially to the overall benefits provided to these species.

7.3.1.2.3 Maryland Forest Conservation Act

Of particular importance in natural resources planning in Maryland is the Forest Conservation Act.[13] The act has its roots in the local tree protection ordinances established by many Maryland counties and local jurisdictions in the 1970s and 1980s. But, it is more encompassing than a simple permitting program for removing and planting trees. It requires developers to identify, map, and characterize forested and other terrestrial habitats on proposed development sites using a process termed forest stand delineation (FSD) and evaluate possible site plan changes and site management practices to minimize impacts to those habitats by developing a forest conservation plan (FCP).[14] The act encourages preservation of existing tree and forest cover before encouraging developers to offset losses by planting new trees. It establishes a high threshold (termed the "breakeven point," measured in acres of forest cover) under which applicants can preserve existing forest cover, especially forest cover in sensitive settings such as wetlands and steep slopes, without having to plant new trees as mitigation. It places exceptionally high emphasis on the avoidance and minimization elements of the traditional environmental mitigation paradigm before directing applicants to compensatory mitigation. Nevertheless, developers who cannot or are unwilling to preserve the requisite amount of forest cover are subject to compensatory mitigation requirements that usually involve planting tree seedlings and saplings in sensitive settings on or close to the project site. As for Section 404 prior to 2008, developers are encouraged to establish forest plantings on site that seek to replicate, as closely as practicable, natural forest cover expected for the landscape setting. Also as for Section 404, forest planting "mitigation banks" sell credits in parts of the state for developers unable to meet their mitigation requirements on site.

The state has issued a technical manual that outlines specific procedures for conducting FSDs and preparing FCPs.[15] In addition to addressing forest stands, FSDs and FCPs must address a comprehensive suite of ecological landscape features such as streams, wetlands, steep slopes, erodible soils, and even historic and archeological features. It is therefore somewhat parallel to the National Environmental Policy Act (NEPA), but its scope is not limited to federal or even state actions. The FSD is roughly comparable to the "affected environment" portion of an environmental impact statement (EIS), and the FCP is roughly comparable to the "environmental consequences" and "mitigation measures" sections of an EIS. Private developers whose projects lie outside the scope of NEPA are still required to obtain approvals under the Forest Conservation Act. It serves as a type of umbrella for many state natural resource planning processes, including the tidal and nontidal wetland acts. However, the Chesapeake Bay Critical Area is specifically excluded from the act, which defers to similarly rigorous requirements under the Critical Area Act.

Of particular relevance to threatened and endangered species, FSDs have to show locations of "critical habitat" for federal and state-listed species, and FCPs must address protection of those habitats. In the context of the Forest Conservation Act, "critical habitat" is merely habitat used by listed species (state or federal), not areas officially designated under the Endangered Species Act or similar state statutes. FCPs must identify priority areas of existing forest for preservation when practicable; the known or possible occurrence of threatened or endangered species in an area greatly increases the need to identify that area as priority. Other key factors are proximity to streams and wetlands; occurrence on steep slopes, floodplains, erodable soils; and adjacency to large blocks of other forest cover. Incorporation of threatened and endangered species protection into the routine land use planning and approval process for land development is a key advance in the level of protection offered, substantially exceeding that available at the federal level for federal projects. Few states offer the same level of protection under state statutes that directly target endangered species. Again, despite its roots, the Forest Conservation Act is focused on forest cover and habitat protection much more than on individual tree protection. It is arguably the most comprehensive ecological land use planning process in the United States today.

Like the Critical Area Act, the Forest Conservation Act encourages preservation of habitat for forest interior-dwelling birds. Specifically, it encourages preservation or establishment of forest patches of at least 100 acres or strips of forest at least 300 feet wide. Both types of patches seek to provide habitat for forest interior birds, while the latter also seek to provide corridors allowing movement of forest interior birds across the landscape. I contributed to an FSD[16] and FCP[17] in 2008 for a proposed nuclear power plant in Calvert County, Maryland. Much of the effort focused on identifying possible opportunities for preserving forest interior bird habitat while achieving other key state natural resource conservation objectives, such as minimizing encroachment into wetlands and wetland buffers and avoiding habitat for the federally and Maryland-listed Puritan tiger beetle and northeastern beach tiger beetle.

Anyone visiting the Maryland suburbs of Baltimore or Washington, D.C., will likely notice the small plots of tree cover that intersperse the office parks and mixed use developments in places where, in other states, one might expect mowed lawn or additional parking lot area. They will also notice how many of the newer suburbs tend to be higher density than comparable suburbs in other states; this reflects Maryland's statewide "Smart Growth" policy of concentrating new land development close to transportation corridors and other public services while discouraging low-density development in rural areas lacking those services. The high-density development encouraged under Smart Growth and preservation of interspersed forest patches encouraged by the Forest Conservation Act do conflict; however, many developers have successfully incorporated preserved clusters of native trees into new high-density development that lend an unexpected "established" aesthetic to those areas. With respect to new areas of low-density development, anyone visiting the more rural outer fringes of Maryland suburbs will likely notice the grids of tree saplings planted as mitigation under the act. Whether the forest preservation and planting efforts fostered by the Forest Conservation Act successfully benefit endangered species will become evident in future decades.

7.1.3.2.4 Maryland Environmental Policy Act

Maryland has a state-level act that closely parallels the National Environmental Policy Act but is directed to state-level projects.[18] The act is one of several "mini-NEPAs" established by several states, mostly northeastern and West Coast states, which have tried to apply the principles of NEPA to the regulation of state actions. As for NEPA and federal projects, the state act does not apply to activities outside the purview of state agencies. It calls for state agencies to prepare "environmental effects reports" patterned very closely after the EIS under NEPA. Because it does not extend to private-sector land development projects, the Maryland Environmental Policy Act is less sweeping than either the Forest Conservation Act or Chesapeake Bay Critical Area Act, although it adds another level of public participation to state projects.

7.3.2 Florida

Florida is perhaps the most biologically unique state in the eastern United States. From north to south, natural habitats in Florida range from temperate deciduous and pine forests in the north that are typical of much of the southeastern coastal plain to tropical evergreen broad-leaved forests in the south that more closely resemble vegetation in the Caribbean. Just as many of Maryland's environmental regulations focus directly or indirectly on the Chesapeake Bay, Florida's environmental regulatory programs focus heavily, directly or indirectly, on protection of the Everglades, the extensive grassy marshes occupying much of south Florida. Just as the Chesapeake Bay is a strong driver of popular will for environmental protection in Maryland, the Everglades are a strong driver of popular will for environmental protection in Florida. The Everglades are a globally unique ecosystem characterized by broad, shallow, slowly running water flowing southward through vast grassy marshes. These marshes are sometimes referred to as the "river of grass," and the frequent small patches of tropical forest on isolated rises amidst the marshes are referred to as tree "islands."

Five key facts help to provide an understanding of the unique ecological habitats of Florida and many of the unusual animal and plant species occupying those habitats:

1. *Most of Florida is a long peninsula.* Ecologists generalize that the flora and fauna of peninsulas tend to become more differentiated from the mainland the farther one moves away from the mainland. The flora and fauna of a peninsula also tend to become increasingly depauperate (less diverse) with increasing distance from the attachment to the mainland. Natural habitats in northern Florida closely resemble habitats elsewhere on the southeastern coastal plain. Natural habitats in extreme southern Florida and the keys bear little resemblance.

2. *Most of Florida is underlain by shallow limestone.* The limestone under the Florida peninsula is a particularly porous type termed oolite. Oolite has a very high transmissivity, that is, water moves through it very rapidly. Precipitation falling in most of Florida leaches (moves downward) very rapidly from the ground surface to the underlying groundwater. The water table is shallow, only a few inches from the surface, over much of Florida, resulting in vast wetlands and numerous freshwater springs. Tannins and other acids in decaying leaves dissolve oolite, resulting in localized depressions that, where shallow, form circular and elliptical wetlands termed cypress domes (usually dominated by pond cypress [*Taxodium ascendens*]) and, where deep, form lakes.

3. *Most of Florida was historically subject to frequent wildfire.* The rapid leaching of rainfall from the ground surface leaves the surface drier than might be expected considering the high annual precipitation. In addition, Florida's humid and hot climate makes thunderstorms especially frequent. The Florida peninsula receives more lightning strikes than anywhere else in the United States. Until modern fire suppression efforts, these frequent lightning strikes sparked wildfires that moved fast over the dry soil surface. The frequent fires prevented the buildup of high fuel loads of leaf duff and fallen limbs, allowing the light fires to sweep over the ground but generally spare the tree canopy. Many of the most familiar vegetation types in Florida, including longleaf pine (*Pinus palustris*) and slash pine (*Pinus elliottii*) forest and even wetland habitats such as cypress domes and the sawgrass (*Cladium jamaicense*) marsh characteristic of the Everglades, are actually fire-dependent successional seres that without fire will progress through succession toward hardwood forests. Natural resource managers in Florida now conduct frequent controlled burns to maintain the presence of these vegetation types.

4. The central part of the Florida peninsula contains sandy ridges. Although many think of Florida as a very flat landscape, the interior of the northern and central parts of the Florida peninsula are actually quite hilly. The ridge-like hills are termed sandhills. To a casual observer, some of these interior areas look more like the hilly piedmont of the eastern United States than the flatter coastal plain.

Where not cleared for agriculture or urban development, the sand-hills support a very dry (xeric) vegetation characteristically dominated by widely spaced longleaf pine trees with a dense understory of turkey oak (*Quercus laevis*).

5. The Florida peninsula extends from the temperate to tropical zones. Although lacking rain forests, the evergreen broadleaved forests of south Florida, including the string of islands at the southern tip termed the Florida Keys, are lush and dense with a large number of different species.

Florida's economy is heavily dependent on tourism. Despite its history of real estate booms and land speculation, people move to Florida for its climate and natural beauty and for recreational pursuits such as bird-watching and fishing that depend on careful natural resource management. People move to Florida, at least in part, for its natural beauty, but in doing so those people displace the very natural features they seek. Many of the large housing developments bear the names of what they replace: Ibis Cove, Old Cypress, Heron Landing, and so on. For many years, the Florida Department of Environmental Protection used a motto "More Protection, Less Process," reflecting a desire to minimize alterations to Florida's uniquely beloved environmental features while minimizing impediments to Florida's economic growth.

7.3.2.1 Statutes and Regulations

Florida protects state-listed threatened and endangered fish and wildlife through the Florida Endangered Species Act.[19] The statute lays out a declaration of policy to protect and conserve Florida's exceptional biological diversity. It states:

> The Legislature recognizes that the State of Florida harbors a wide diversity of fish and wildlife and that it is the policy of this state to conserve and wisely manage these resources, with particular attention to those species defined by the Fish and Wildlife Conservation Commission, the Department of Environmental Protection, or the United States Department of Interior, or successor agencies, as being endangered or threatened. As Florida has more endangered and threatened species than any other continental state, it is the intent of the Legislature to provide for research and management to conserve and protect these species as a natural resource.[20]

This declaration of policy is further described in regulations issued under the act, as follows:

> The purpose and intent of this rule chapter, in concert with an objective that lawful nature-based recreational activities may be managed

to be compatible with such species protection measures, is to conserve or improve the status of endangered and threatened species in Florida to effectively reduce the risk of extinction through the use of a science-informed process that is objective and quantifiable, that accurately identifies endangered and threatened species that are in need of special actions to prevent further imperilment, that identifies a framework for developing management strategies and interventions to reduce threats causing imperilment, and that will prevent species from being threatened to such an extent that they become regulated and managed under the federal Endangered Species Act of 1973, as amended.[21]

The regulations emphasize a practical approach seeking to balance the goals of conservation with the recreational activities that are a key part of Florida's quality of life and tourism that drives much of the state's economy. The regulations also establish a preventive tone, seeking to conserve species through actions guided by the state before populations of rare species decline to the point that species become federally listed and subject to possibly more rigorous protection under the Federal Endangered Species Act.

The Florida Endangered Species Act defines endangered and threatened species in a manner generally similar to the federal definitions in the Endangered Species Act. The state defines endangered species as "any species of fish and wildlife naturally occurring in Florida, whose prospects of survival are in jeopardy due to modification or loss of habitat; overutilization for commercial, sporting, scientific, or educational purposes; disease; predation; inadequacy of regulatory mechanisms; or other natural or manmade factors affecting its continued existence."[22] The state defines threatened species as "any species of fish and wildlife naturally occurring in Florida which may not be in immediate danger of extinction, but which exists in such small populations as to become endangered if it is subjected to increased stress as a result of further modification of its environment."[23] The substance of the state definitions is the same as that of the federal definitions, except for the focus on the continued presence of species in Florida rather than global extinction. The Florida definition for endangered incorporates specific scientific criteria for meeting the definition. These criteria closely resemble federal criteria for listing as endangered.

State-listed species other than plants are regulated by the Florida Fish and Wildlife Conservation Commission.[24] The commission limits state-endangered species to federally listed endangered species occurring in Florida.[25] It does not designate species as state endangered that are not also federally listed as endangered. While also recognizing federally listed threatened species as state threatened, the commission also designates other species not federally listed as state threatened. The commission's regulations include detailed criteria for designating species as state threatened.[26]

As of October 2011, the commission officially recognized 131 state species in the preceding categories, including 46 federally listed endangered species, 19 federally listed threatened species, one federally listed threatened species based on similarity of appearance, one federally designated extinct species, 21 state-threatened species, and 43 state species of special concern.[27] The species of special concern are not subject to regulatory protection against take. Table 7.1 presents a more detailed breakout.

State-listed plant species in Florida are designated by the Florida Department of Agriculture and Consumer Services. As of 2010, the department listed 440 endangered plant species and 117 threatened plant species.[28] Not all of the state endangered or state threatened plant species recognized by the department are federally listed. Unlike the Florida Fish and Wildlife Conservation Commission, the department does not limit state endangered species to federally listed endangered species. In fact, the state lists are considerably more expansive than the federally listed species known to occur in Florida. The department also recognizes eight "commercially exploited" plant species whose populations have been compromised by collectors or other forms of commercial exploitation. The department has established an Endangered Plant Advisory Council consisting of seven professionals who meet at least once per year to consider changes to the list.[29]

Florida defines take and incidental take in a manner quite similar to the federal definitions.

> (4) Take—to harass, harm, pursue, hunt, shoot, wound, kill, trap, capture, or collect, or to attempt to engage in such conduct. The term "harm" in the definition of take means an act which actually kills or injures fish or wildlife. Such act may include significant habitat modification or degradation where it actually kills or injures wildlife by significantly impairing essential behavioral patterns, including breeding, feeding or sheltering. The term "harass" in the definition of take means an intentional or negligent act or omission which creates the likelihood of injury to wildlife by annoying it to such an extent as to significantly disrupt normal behavioral patterns which include, but are not limited to, breeding, feeding or sheltering.[30]
>
> (5) Incidental take—any taking otherwise prohibited, if such taking is incidental to, and not the purpose of the carrying out of an otherwise lawful activity.[31]

Florida has established permitting programs for both take and incidental take. Regarding incidental take of state-listed threatened species, the Florida regulations state:

> The Commission may issue permits authorizing incidental take of State-designated Threatened species upon a conclusion that the following permitting standards have been met: the standards for species when contained in Rule 68A-27.003, F.A.C., take precedence; for blackmouth shiner, striped mud turtle, Florida mastiff bat, and pillar coral, a permit may be issued if the permitted activity clearly enhances the survival

potential of the species; for all other State-designated Threatened species, the permit may be issued when there is a scientific or conservation
benefit and only upon a showing by the applicant that the permitted
activity will not have a negative impact on the survival potential of the
species. Factors which shall be considered in determining whether a
permit may be granted are:

1. The objectives of a federal recovery plan or a state management
   plan for the species sought to be taken;
2. The foreseeable long range impact over time if take of the species is authorized;
3. The impacts to other fish and wildlife species if take is authorized;
4. The extent of injury, harm or loss of the species;
5. Whether the incidental take could reasonably be avoided, mini-
   mized or mitigated by the permit applicant;
6. Human safety; and
7. Other factors relevant to the conservation and management of
   the species.[32]

Note that the Florida regulations establish an incidental take permit process
for state threatened species; incidental take of state endangered species is not
sanctioned. The Florida regulations specifically exempt conservation land
management practices, agriculture (if conducted in accordance with BMPs),
and fire suppression activities from the need for incidental take permits.[33]

One Florida state-listed species that plays a key role in planning many
land development projects is the gopher tortoise (*Gopherus polyphemus*),
which is designated by the state as threatened. The gopher tortoise population in Florida is not federally listed and receives no protection under the
Endangered Species Act.[34] Gopher tortoises are large, long-lived reptiles that
den in underground burrows that they dig in sandy uplands in Florida's
sandhill forests and drier flatwoods, as well as dry sandy pastures and
yards (that likely supported sandhill forests prior to development). They dig
deep burrows for shelter and share the burrows with more than 350 other
species.[35] The species does not favor Florida's extensive wetland habitats. It
therefore tends to lie in the path of development projects that are preferentially directed to uplands rather than adjoining uplands. However, Florida
requires developers to survey uplands on development sites for gopher tortoise burrows, avoid areas of burrows if possible, and relocate other gopher
tortoises to safe habitat.

Even though the gopher tortoise is not federally listed, the state effort to
direct development away from gopher tortoise burrows indirectly helps to
protect a federally listed threatened species, the indigo snake (*Drymarchon
corais couperi*). The indigo snake uses gopher tortoise burrows as homes.
The snake was listed as threatened in 1978 because of habitat loss, collection
for the pet trade, and impacts from "rattlesnake roundups" where various
snakes are pursued for sport and to rid areas of snakes feared by some people.[36] The upland habitats favored by indigo snakes are not subject to Section

404 of the Clean Water Act. Developments affecting only those habitats, and no wetlands, do not require federal permits under Section 404 or other federal regulations. There is therefore no opportunity for the U.S. FWS to review such projects for incidental take under Section 7 of the Endangered Species Act. However, the avoidance measures that might have been recommended by the FWS still result from the state procedures. This is an excellent example of state protection of endangered species complementing federal protection.

Many of Florida's state-listed species are wading birds. Most wading birds are large, colorful species that are easily viewed and enjoyed by nonspecialists without binoculars or other equipment. They are also highly photogenic. Perhaps the best-known southern Florida wading bird, the pink flamingo, actually the American flamingo (*Phoenicopterus ruber*), is infrequent in the wild in Florida, and most, if not all, individuals in Florida may descend from escapes from zoos. A far more frequent and just as spectacular pink wading bird that is actually native to Florida is the roseate spoonbill (*Platalea ajaja*), listed by Florida as a species of special state concern. Most pink birds in Florida are roseate spoonbills. The great blue heron (*Ardea herodias*), a very large wading bird actually common over much of the United States, is also very common in Florida, but southern Florida, especially the keys, is also inhabited by the rare white morph of the great blue heron, referred to as the great white heron. Numerous uninhabited islands in the Florida Keys are preserved as the Great White Heron National Wildlife Refuge. Some other rarely seen wading birds include the reddish egret (*Egretta rufescens*), generally limited to beaches in Texas and southern Florida; the limpkin (*Aramus guarauna*), generally limited to wetlands in interior southern Florida; and the anhinga (*Anhinga anhinga*), a black snake-like bird locally common in the freshwater wetlands of Florida but rare or absent elsewhere.

Most wading bird species of Florida were quite common until the early twentieth century. The colorful plumage of the birds became popular to accessorize ladies' hats in fashion at that time. Hunters moving to Florida were paid enormous amounts of money for the feathers, such that the value per ounce of features reportedly exceeded at the time the value per ounce of gold. The result was a "gold rush" on wading bird plumage that rapidly depleted the populations of many species to very low levels. Ironically, the small outpost town of Flamingo that presently serves as a popular bird-watching spot in Everglades National Park was once a base camp for plume hunters. The realization of severely depleted wading bird populations led to the establishment of hunting bans somewhat reminiscent of the future Endangered Species Act. Florida enacted a state law in 1901 banning the hunting for the plumage of wading birds during their breeding season. The National Audubon Society, a leading conservation group that protects and manages many of the nation's premier bird-watching sites, was founded in part in response to the decline of wading birds due to the plume hunters.[37] The populations of most wading bird species targeted by the plume hunters have recovered to levels secure from possible extinction, although the stresses formerly imposed by the plume hunters have been replaced in part by stresses related to draining and development of Florida's wetland habitats. The recovery of most of Florida's wading bird populations from populations early in the twentieth century that would have warranted listing under the Endangered Species Act (had it existed then) to stable although still reduced levels today is an encouraging sign that the protections advocated under the act can at least sometimes be effective.

The only federally listed wading bird in Florida is the endangered wood stork (*Mycteria americana*). The wood stork is a visually imposing bird, over five feet in height with a bizarre black, vulture-like head. Wood storks tend to breed at rookery sites situated in trees surrounded by broad areas of herbaceous wetland habitat that they use for foraging. They tend to return to the same breeding sites in successive years but have abandoned those sites after the associated wetlands have been drained for agriculture or development. Wood stork populations declined alarmingly in the middle to late twentieth century, attributed mostly to wetland drainage and cutting of cypress swamps.[38]

For an endangered species, wood storks have been relatively easy to find in parts of Florida, especially southwest Florida. I have observed wood storks frequently foraging in the ubiquitous roadside drainage ditches in developed areas in southwest Florida, much more so than other unlisted wading birds such as the limpkin or reddish egret. The FWS initiated a review in 2010 to evaluate whether down-listing from endangered to threatened is warranted.[39] However, 2012 marks the fifth time in the last six years that wood storks have not nested at their largest known nesting site, Corkscrew Sanctuary in southwestern Florida, around which substantial development has occurred in recent years.[40] Recovery of endangered species can be both fleeting and illusional; this may at least partially explain the reticence of many scientists to recommend confidently that species be delisted, despite the ongoing political pressure to demonstrate the "success" of endangered species regulations.

7.3.2.2 Other Related Florida Statutes

As in Maryland and many other states, conservation of threatened and endangered species is intimately tied to conservation of wetlands. Wetlands have long been a quandary in Florida. Wetlands are more extensive than nonwetlands (uplands) in many parts of Florida. Almost no development can occur in many parts of Florida without affecting wetlands—often lots of wetlands. Whereas wetland delineation reports more commonly map small areas of wetlands amid large areas of uplands, many wetland delineations in Florida map small areas of uplands amid large areas of wetlands. Designing developments to avoid encroachment into the few areas of wetlands on a site is a realistic objective for the former situation; it is impossible in the latter situation. Whereas wetland mitigation in many areas of the country focus on creating wetlands out of uplands or restoring wetlands formerly drained or filled for past activities, wetland mitigation in Florida largely focuses on preserving wetlands, especially the highest-quality remaining wetlands, including wetlands recognized as most beneficial to endangered species such as the wood stork.

Perhaps what is most unique about state-level wetlands protection in Florida is how wetlands are defined and delineated.[41] Wetlands are defined at the state level in Florida as

> those areas that are inundated or saturated by surface water or ground water at a frequency and a duration sufficient to support, and under normal circumstances do support, a prevalence of vegetation typically adapted for life in saturated soils. Soils present in wetlands generally are classified as hydric or alluvial, or possess characteristics that are associated with reducing soil conditions. The prevalent vegetation in wetlands generally consists of facultative or obligate hydrophytic macrophytes that are typically adapted to areas having soil conditions described above. These species, due to morphological, physiological, or reproductive adaptations, have the ability to grow, reproduce or persist in aquatic environments or anaerobic soil conditions. Florida wetlands generally include swamps, marshes, bayheads, bogs, cypress domes and strands, sloughs, wet prairies, riverine swamps and marshes, mangrove swamps and other similar areas. Florida wetlands generally do not include longleaf or slash pine flatwoods with an understory dominated by saw palmetto.[42]

The Florida definition begins with verbiage very similar to the federal definition of wetlands used for Section 404.[43] But, then it introduces elements targeted to Florida's unique landscape. Part of the Florida wetland delineation process is the need for using "reasonable scientific judgment," described as involving

> the ability to collect and analyze information using technical knowledge, and personal skills and experience to serve as a basis for decision making. Examples of situations where reasonable scientific judgement [sic] is very important include: ecotonal, seasonally wet or occasionally wet lands which are not the wetlands intended by the statutory definition, wetland communities dominated by non-listed plant species such as *Quercus virginiana* (live oak) and *Magnolia grandiflora* (southern magnolia), i.e., hydric hammock, altered areas which still have relict wetland vegetation and/or hydric soils but may have lost the hydrology necessary to maintain a wetland condition, and wetland ecotones, especially throughout south Florida.[44]

Like the federal wetland delineation methodology, the Florida methodology relies on the three parameters of vegetation, soils, and hydrology. But, the plant species constituting wetland vegetation are more limited in Florida. Most notably, Florida does not recognize lands dominated by slash pine (*Pinus elliottii*) with an understory of saw palmetto (*Serenoa repens*) as wetlands. Although this might seem to be a fine distinction, it excludes from wetland regulation much of Florida's pine flatwoods, a pine-dominated plant community occupying large areas of flat, somewhat poorly drained soils forming a landscape matrix separating distinct wetland features such as cypress swamps, river bottom swamps, bay and gum swamps, and grass-dominated wetlands commonly referred to as wet prairies. Even though many of Florida's pine flatwoods have been cleared for agriculture and forestry, many former flatwood areas presently support managed pine stands undergrown by saw palmetto, also excluded from state wetland regulation by this definition.

One key element of public opposition to Section 404, as well as many other environmental regulatory programs, including those of the Endangered Species Act, is that they are not technically suited for all landscapes. The Florida definition appears to be a commonsense approach to developing a workable state-level regulatory program to protect wetlands in a landscape such as Florida, where so much of the landscape is either wetland or displays wetland-like properties. Perhaps considering its conservative history and economic dependence on development coupled with the linkage of that development to the natural aesthetics that attract people to the state, Florida has done more than perhaps any other state to reconcile the need for continued economic growth with protecting the resources that underlie that very growth. It must be borne in mind, however, that Florida's wetland definition, no matter how much better suited it is to Florida's landscape, is still applicable only to Florida's statutes; the federal wetland definition in 33 C.F.R. 328.3 is the only definition recognized for purposes of Section 404 of the Clean Water Act, even in Florida.

Perhaps as an outgrowth of this reconciliation effort, Florida has been a leader in using functional assessment as a means to funnel wetland preservation and regulation efforts to those wetlands most deserving of the efforts. Florida presently requires the use of a functional assessment method developed specifically for use in Florida, the unified mitigation assessment method (UMAM). Over 40 wetland functional assessment methods have been developed for use in the United States; what makes UMAM unusual is that it has been established by regulation,[45] and its use is mandatory by the water management districts when reviewing applications for environmental resource permits involving wetland impacts in Florida. UMAM provides a basis for translating wetland acreage into functional units; wetland impacts traditionally expressed as acres lost can be expressed using UMAM as functional units lost. Mitigation credits can likewise be expressed as UMAM functional units. The Florida regulation establishing UMAM throughout Florida states the following:

> The methodology in this Chapter provides a standardized procedure for assessing the functions provided by wetlands and other surface waters, the amount that those functions are reduced by a proposed

impact, and the amount of mitigation necessary to offset that loss. It does not assess whether the adverse impact meets other criteria for issuance of a permit, nor the extent that such impacts may be approved. This rule supersedes existing ratio guidelines or requirements concerning the amount of mitigation required to offset an impact to wetlands or other surface waters.[46]

7.3.3 Texas

Texas, the largest state in terms of land area within the conterminous 48 states, encompasses an usually diverse spectrum of ecological conditions, ranging from various pine and deciduous forest habitats typical of parts of the American Southeast in its eastern parts of the states; various grassland and savanna habitats in the central and northern parts of the state; to various semidesert, desert, and montane habitats in the western parts of the state. The Texas Parks and Wildlife Commission (TPWC) recognizes 10 "natural regions," including (in rough order from east to west) piney woods, gulf coast prairies and marshes, oak woods and prairies, blackland prairies, South Texas brush country, Edwards Plateau, Llano Uplift, rolling plains, high plains, and Trans-Pecos.[47] The names of these natural regions also serve as good descriptive summaries. The EPA, reporting to the Texas Commission on Environmental Quality, recognizes 10 distinct ecoregions, generally but not exactly corresponding to the aforementioned natural regions, as follows:

- Arizona/New Mexico mountains;
- Chihuahuan deserts;
- High Plains;
- Southwestern tablelands;
- Central Great Plains;
- Cross Timbers;
- Edwards Plateau;
- Southern Texas plains;
- Texas blackland prairies;
- East central Texas plains;
- Western Gulf coastal plain; and
- South central plains.[48]

The EPA defines ecoregions as "areas within which there is similarity in the mosaic of biotic and abiotic components of both terrestrial and aquatic ecosystems."[49] The Nature Conservancy describes Texas as one of four states (along with California, Hawaii, and Alabama) having "exceptional levels of biodiversity."[50] It states:

Looming large in both popular imagination and in biological diversity, Texas ranks highly in diversity, endemism, and number of extinctions. Occupying a central position along the nation's southern border, this vast state overlaps several major ecological regions, including the southwestern deserts, the Great Plains, the humid Gulf Coast, and, at the state's southern tip, the Mexican subtropics. As a result, many species reach distributional limits in Texas, and a strange blend of eastern and western species commingle within the state. Certain unusual landforms contribute to the state's high rankings, including the Edwards Plateau, a limestone region that supports some of the rarest species in the nation.[51]

The TPWC maintains two useful databases on known occurrences of federal- and Texas-listed rare, threatened, and endangered species and sensitive natural habitats: county lists and the Texas Natural Diversity Database (TxNDD). The former provides lists of known and potentially occurring listed species by Texas county; the latter compiles available data for indicated U.S. Geological Survey 7.5-minute topographic quadrangles. The former is available online with information available immediately; the latter requires submission of a written request with responses in about five business days. TPWC recommends consulting both information sources in the early stages of project planning. Once the project is at a conceptual design stage, TPWC offers an individualized review termed the Wildlife Habitat Assessment Program (WHAB) review. Project proponents must apply in writing for a WHAB review by submitting a form with project description information. Responses can take four to six weeks.[52]

7.3.3.1 Statutes and Regulations

The TPWC establishes state lists of endangered and threatened species and administers state permits for impacts to those species under the Texas Endangered Species Act. Texas defines species of fish and wildlife as endangered if they are on "the United States List of Endangered Fish and Wildlife" or "on the list of fish and wildlife threatened with statewide extinction as filed by the director of the department."[53] Texas defines endangered fish and wildlife only; it does not define (or regulate) endangered plants or insects. The statute defines fish and wildlife as "any wild mammal, aquatic animal, wild bird, amphibian, reptile, mollusk, or crustacean, or any part, product, egg, or offspring, of any of these, dead or alive."[54] The use of the term *statewide extinction* is technically incorrect; what it means is *extirpation* from the state. Texas defines a threatened species as "any species that the department has determined is likely to become endangered in the future."[55]

As a state with a long tradition of individual freedoms and property rights, the Texas statute is set up to be readily responsive to the will of the people. The statute contains a provision allowing as few as three persons to petition the state to add or delete species to (or from) the state list.[56] The statute establishes a permit requirement to "possess, take, or transport" endangered fish or wildlife.[57] But, as might be expected, the statute has no language implying regulation of incidental take. Of course, the statutory language of the Endangered Species Act itself does not specifically address incidental take; any regulation of incidental take at the federal level is a product of interpretation. Otherwise, the actions regulated under Texas statute do not greatly differ from those regulated by the federal government or by many other states.

Texas prohibits the following with respect to state-listed endangered and threatened species without a permit from TPWC:

(1) taking, possessing, propagating, transporting, exporting, selling or offering for sale, or shipping any species of fish or wildlife listed by the department as endangered;
(2) taking, possessing, propagating, transporting, importing, exporting, selling, or offering for sale any species of fish or wildlife listed in this subchapter as threatened; or
(3) selling or propagating for sale any species of fish or wildlife listed by the department as endangered.[58]

Unlike the federal regulations developed by the FWS for the Endangered Species Act, the Texas state-level regulations do not define take. Incidental take, however, is widely recognized as not prohibited under the Texas regulations.

7.3.3.2 Other Related Texas Statutes

Unlike Florida and Maryland, but like most southern, midwestern, and non-coastal western states, Texas does not specifically regulate wetland impacts at the state level. However, like most states without direct wetland protection statutes, the Texas Commission on Environmental Quality (TCEQ) retains the ability to regulate wetland impacts indirectly through the Section 401 Water Quality Certification process under the federal Clean Water Act.[59] Texas's regulations for implementing Section 401[60] establish a policy "to achieve no overall net loss of the existing wetlands resource base with respect to wetlands functions and values in the State of Texas."[61]

This policy is, at least on the surface, similar to policies set for no net loss of wetlands by many northeastern and West Coast states known for more aggressive wetland regulation. However, unlike many of those states, Texas does not require applicants whose wetland impacts qualify under general permits, including nationwide general permits, to seek additional review with respect to wetlands.[62] In addition, impacts to wetlands deemed to be isolated rather than adjacent with respect to 33 C.F.R. 328, as a result of two Supreme Court cases in 2001 and 2006, respectively, are excluded from Section 401 review. That exclusion was not established by the state of Texas; Section 401 can only be used by states to review Section 404 applications covering impacts to areas specifically within the scope of the Clean Water Act.

In contrast, Maryland, Florida, and some other states with their own wetland protection statutes, independent of Section 401, still apply to some or all of the wetlands excluded from Section 404 coverage by the subject Supreme Court decisions.

Furthermore, TCEQ has developed two tiers of review for state water quality certification for individual Section 404 permits, that is, those not qualifying under a general permit. Projects affecting less than three acres of wetlands or 1,500 linear feet of stream channel (or a combination of more than three acres of wetlands where each 500 linear feet of stream channel impact is counted as one acre of wetland impact) are termed Tier I projects. If applicants complete a checklist documenting the incorporation of BMPs into the design, TCEQ does not perform additional review. Other projects are deemed to be Tier II projects and receive a more comprehensive review by TCEQ. TCEQ specifically excludes specific high-value wetlands—such as pitcher plant bogs, cypress and gum swamps, mangrove swamps, coastal dune swales, and Caddo Lake (a Ramsar wetland of international importance)—from the Tier I designation (thereby requiring a Tier II review despite falling below the impact area threshold).[63] This exclusion encompasses many wetlands especially likely to harbor threatened and endangered species.

Notes

1. U.S. Department of Agriculture. 2011. PLANTS Profile for American Lotus (*Nelundo lutea*). Available at http://plants.usda.gov/java/profile?symbol=NELU. Accessed September 23, 2011.
2. Braun, E.L. 2001. *Deciduous Forests of Eastern North America*, reprint of first edition, 1950, Blackburn Press, Caldwell, NJ, 506 pp.
3. Maryland Natural Resources article. 10-2A-02.
4. COMAR 08.03.08.
5. COMAR 08.03.08.01(14).
6. COMAR 08.03.08.09.
7. COMAR 08.03.08.04 and 08.03.08.07.
8. Maryland Natural Resource Article 8-1801 et seq.
9. COMAR 27.01.01 to 27.03.01.
10. Chesapeake Bay Critical Area Commission. 1986. *A Guide to the Conservation of Forest Interior Dwelling Birds in the Critical Area*, Guidance Paper No. 1, July, 13 pp. plus appendix. Annapolis, MD.
11. Askins, R.A. 2002. *Restoring North America's Birds, Lessons from Landscape Ecology*, 2nd edition, Donnelly, Harrisonville, VA, 332 pp.
12. U.S. Fish and Wildlife Service. 2006. News Release: U.S. Fish and Wildlife Service Finds Cerulean Warbler Not Warranted for Endangered Species Act Listing. December 6; last updated May 9, 2011. Available at http://www.fws.gov/midwest/eco_serv/soc/birds/cerw/cerw12mnthfindnr.html. Accessed March 4, 2012.

13. Maryland Natural Resources Article 5-1601–1613.
14. COMAR 08.19.09.
15. Howell, G.P., and Ericson, T. 1997. *State Forest Conservation Technical Manual*, 3rd edition, Maryland Department of Natural Resources, Annapolis, MD.
16. Tetra Tech NUS. 2008. *Forest Stand Delineation Report for Proposed Calvert Cliffs Nuclear Power Plant Unit 3 Project Area, Calvert Cliffs Nuclear Power Plant Site, Calvert County, Maryland*. 15 pp. plus appendices. Germantown, MD.
17. Tetra Tech NUS. 2008. *Forest Conservation Plan for Proposed Calvert Cliffs Nuclear Power Plant Unit 3 Project Area, Calvert Cliffs Nuclear Power Plant Site, Calvert County, Maryland*. 12 pp. plus appendices. Germantown, MD.
18. Maryland Natural Resources Article 1-301–305.
19. Florida Statutes Annotated, Title XXVIII, 379.2291.
20. Ibid., 379.2291(2).
21. F.A.C. 68A-27.0001.
22. Florida Statutes Annotated, Title XXVIII, 379.2291(3)(b).
23. Ibid., 379.2291(3)(c).
24. 68A-27 F.A.C.
25. 68A-27.001(1)(a) F.A.C.
26. 68A-27.001(3) F.A.C.
27. Florida Fish and Wildlife Conservation Commission. 2011. *Florida's Endangered and Threatened Species*. October, 10 pp. Tallahassee, FL.
28. Florida Department of Agriculture and Consumer Services. 2010. *Notes on Florida's Endangered and Threatened Plants*. 112 pp. Tallahassee, FL.
29. Ibid.
30. F.A.C. 68A-27.0001(4).
31. F.A.C. 68A-27.0001(5).
32. F.A.C. 68A-27.0007(2)(b).
33. F.A.C. 68A-27.0007(2)(c–e).
34. Gopher tortoise populations in some parts of the country outside Florida are federally listed.
35. Florida Fish and Wildlife Conservation Commission. Gopher Tortoise. Available at http://myfwc.com/wildlifehabitats/managed/gopher-tortoise/. Accessed January 2, 2012.
36. U.S. Fish and Wildlife Service. 1978. *Final Determination of Threatened Status for Eastern Indigo Snake*, 43 FR 4026-4029, January 31.
37. National Audubon Society. 2012. History of Audubon and Waterbird Conservation. Available at http://birds.audubon.org/history-audubon-and-waterbird-conservation. Accessed January 3, 2012.
38. U.S. Fish and Wildlife Service. 1996. *Revised Recovery Plan for the U.S. Breeding Population of the Wood Stork*. U.S. Fish and Wildlife Service, Atlanta, GA, 41 pp.
39. *Federal Register,* 75(182), 57426–57431 (September 21, 2010).
40. *Miami Herald.* 2012. Wood Storks Shun Corkscrew Swamp Sanctuary. March 5. Available at http://www.miamiherald.com/2012/02/25/2660035/wood-storks-shun-corkscrew-swamp.html. Accessed March 5, 2012.
41. Chapter 62-340, Florida Administrative Code, *Delineation of the Landward Extent of Wetlands and Surface Waters.*
42. 373.019(17), Florida Statutes.
43. 33 C.F.R. 328.3.

44. Florida Department of Environmental Protection. The Florida Wetlands Delineation Manual. Available at http://www.dep.state.fl.us/water/wetlands/delineation/docs/intro.pdf.
45. Chapter 62-345, Florida Administrative Code, *Universal Mitigation Assessment Method.*
46. Chapter 62-345.100(2).
47. Texas Parks and Wildlife Commission. Endangered and Threatened Species. Available at http://www.tpwd.state.tx.us/huntwild/wild/species/endang/. Accessed March 2, 2012. *Natural Region Map. Based on Preserving Texas' Natural Heritage*, LBJ School of Public Affairs Policy Research Project Report 31, 1978.
48. Griffith, G.E., Bryce, S.A., Omernik, J.M., Comstock, J.A., Rogers, A.C., Harrison, B., Hatch, S.L., and Bezanson, D. 2004. *Ecoregions of Texas* [color poster with map, descriptive text, and photographs], U.S. Geological Survey, Reston, VA (map scale 1:2,500,000), Figure 1, p. v.
49. Ibid., p. 1 (Introduction).
50. Stein, Bruce A. 2002. *States of the Union: Ranking America's Biodiversity*, NatureServe, Arlington, VA, p. 2
51. Ibid., p. 8.
52. Texas Parks and Wildlife Commission. State of Texas Threatened and Endangered Species Regulations. Available at http://www.tpwd.state.tx.us/huntwild/wild/species/endang/regulations/texas/index.phtml. Accessed March 4, 2012.
53. Texas PW Code Ann. 68.002.
54. Texas PW Code Ann. 68.001.
55. Section 65.175 T.A.C.
56. Texas PW Code Ann. 68.005.
57. Texas PW Code Ann. 68.006.
58. Section 65.171(b) T.A.C.
59. 33 U.S.C. 1341
60. 30 T.A.C. 279.1-279.13.
61. 30 T.A.C 279.2(b).
62. 30 T.A.C. 279.12.
63. Texas Commission on Environmental Quality. 2004. State Water Quality Certification of Section 404 Permits, April 12.

8

Future of the Endangered Species Act

8.1 Introduction

The first seven chapters of this book have been largely factual, examining the history of, the regulatory content of, and the scientific basis for the Endangered Species Act. It is hoped these chapters will help environmental consultants and other practitioners improve their work in the context of the act. It is also hoped those chapters will give other readers a window into how environmental practitioners perform their professional duties and the challenges they routinely face in carrying out these duties. But, the American environmental regulatory scene has always been dynamic and fast changing. Most readers probably want to know how the Endangered Species Act, now approaching 40 years of implementation, may change in the near and long-term future. Environmental policy, like government policy in most fields, is inherently very unpredictable. But, through a brief high-level discussion of some of the leading controversies over the last several years, it may be possible to prognosticate at least the near-term direction for the act.

Section 8.2 discusses sources of support for the Endangered Species Act. Section 8.3 examines the foundations of opposition to the act. Section 8.4 considers some of the more controversial recent incidences of controversy over the Endangered Species Act and related environmental protection statutes, such as Section 404 of the Clean Water Act. Section 8.5 is a brief summary of how I expect the Endangered Species Act to change over the next several years. Section 8.5 is, admittedly, purely opinion and speculation; it is therefore brief. It represents the opinion of just one of the many environmental practitioners working in the United States today. But, the objective for Section 8.5 is not to publicize the opinions of Peyton Doub; it is to get other readers to formulate their own informed opinions.

8.2 Basic Sources of Support for the Endangered Species Act

The history of the motivations and controversies that led to the initial implementation of the Endangered Species Act is discussed in Chapter 1 and is not repeated here. The following discussion instead waxes a bit more philosophical. The viewpoints presented constitute my perspective. Some readers may disagree, or they may feel that additional or different viewpoints are more appropriate.

John Muir, the founder of the Sierra Club, is frequently quoted as saying, "When we try to pick out anything by itself, we find it hitched to everything else in the Universe."[1] Also Leopold, author of the influential philosophical conservation treatise *A Sand County Almanac*, stated: "To keep every cog and wheel is the first precaution of intelligent tinkering," and "A thing is right when it tends to preserve the integrity, stability and beauty of the biotic community. It is wrong when it tends otherwise."[2] The species we encounter in the fields and forests are part of larger ecosystems. As described previously in Chapter 2, each species present in an ecosystem occupies a unique position termed its niche, and niches are very difficult to describe and delineate. Rendering a species extinct, or even extirpating it from a geographic subarea of its range, can have unpredictable ecological consequences. The classic example is the Tambalacoque tree (*Sideroxylon grandiflorum*), endemic to the island nation of Mauritius in the Indian Ocean, that ceased to reproduce normally after extinction of the dodo bird (*Raphus cucullatus*). The dodo was thought to have primed the tree's seeds for germination by passing the seeds through the bird's digestive system. The dodo went extinct in the seventeenth century, and the relationship between the extinction of the dodo and the decline of the Tambalacoque tree was not postulated until the twentieth century. Although this example of obligate mutualism between a living and an extinct species has been subject to question,[3] it still illustrates how extinctions can have unexpected consequences not evident until years later. How long will it take to fully understand the ramifications of the more recent extinction of species such as the passenger pigeon (*Ectopistes migratorius*), Carolina parakeet (*Conuropsis carolinesis*), or dusky seaside sparrow (*Ammodramus maritimus nigrescens*)?

Furthermore, extinction is permanent; it is irreversible. Society cannot decide that a species is expendable and allow it to go extinct and then bring it back should a future generation desire it back. Although there is persistent speculation that scientists might one day be able to clone endangered and extinct species, this concept largely remains in the realm of science fiction. Researchers reported in 2000 that somatic (nonreproductive) cells from a gaur bull (*Bos gaurus*), a wild ox near extinction, were electrofused with reproductive cells from domestic cows to produce embryos that were successfully gestated for several days before dying, providing evidence that cloning might one day be used to propagate endangered species or regenerate extinct species.[4] Even though European scientists were able to clone an endangered mountain goat species, the Pyrenean ibex (*Capra pyrenaica pyrenaica*), using the womb of a domestic goat (*Capra aegagrus hircus*), the offspring was short lived, and other attempts to clone mammals have experienced unexpected technical challenges.[5] Even if the biotechnological challenges can be overcome, there are ecological challenges of reintroducing the cloned individuals into their former niches. Cloning is not a serious consideration in any recovery plans or other documents produced or decisions made in the context of the Endangered Species Act today.

Considering the concepts discussed, there are several motivations that can underlie a desire to protect threatened and endangered species. They may be broadly classified as follows:

1. *Moral*: Many feel that we, either as human beings or as a nation, have a responsibility to protect species from extinction. As noted, extinction is irreversible; we cannot allow species to go extinct now and decide we want them back later. Any decision to allow a species to go extinct is irrevocably passed to future generations; any species preserved from extinction is bequeathed to future generations— but only until one of those generations lets it go extinct. The moral motivation may, as is true for me, derive from religion—we have a responsibility to God to be good stewards of what He has created— or the motivation may simply derive from a perceived responsibility to nature or to the planet we call home. From this perspective, any extinction is unacceptable. No extinctions can be justified simply because of the effort needed or cost required to prevent extinction. The Endangered Species Act as originally developed is consistent with this moral perspective.

2. *Aesthetic*: People enjoy and appreciate the diversity of life in the natural world. Laws would not likely have been passed to protect wading birds in the early 1900s if the birds were not colorful and beautiful. The excitement over the possible rediscovery of the visually striking ivory-billed woodpecker (*Campephilus principalis*) would not likely have been so intense if a similar situation occurred for a small insect or obscure plant with only small, drab flowers. The passion shared by bird-watchers for the wondrous diversity of birdlife, many of whom gauge the success of their hobby by how many species they can observe over their lifetimes (their "life list") is a key force driving the American public to demand conservation measures from their politicians, including but not limited to protections for endangered species. Although the Endangered Species Act is more suited to the moral perspective, the aesthetic perspective has likely been a strong force driving the American public to establish and retain the act and continue to list more and more species under the act.

3. *Ecological*: As scientists have gained increasing knowledge of the complex interrelations among the species inhabiting natural habitats, they have increasingly been able to argue the need for preserving all of the pieces making up those interrelations. The system is so complex, according to those arguing from the ecological perspective, that we can not fully predict the possible ramifications of allowing any species to go extinct. Those arguing for protecting species from an ecological perspective share a key element with those arguing from the moral perspective: No species are expendable, no extinctions are justifiable.

4. *Practical*: Beyond the more esoteric arguments for conserving species is the argument that human society benefits in tangible ways from the presence of species. Large predators such as gray wolves (*Canis lupus*) and Florida panthers (*Puma concolor coryi*) help keep populations of their prey species, such as whitetail deer (*Odocoileus virginianus*), at manageable levels that do not interfere with human activities. Rare plant species may contain chemicals that might one day serve as beneficial pharmaceuticals for future generations or serve as future crop plants. The practical argument is economic—the costs of conserving endangered species are expected to be offset by the possible benefits offered by those species to future generations.

The Bible includes several passages speaking of the beauty of the innumerable species in the natural world and reminding humankind of its responsibility as stewards of the natural environment. Examples include the following:

- In wisdom you made them all, the earth is full of your creatures. There is the sea, vast and spacious, teeming with creatures beyond number—living things both large and small.[6]
- As for you, my flock … is it not enough for you to feed on good pasture? Must you also trample the rest of your pasture with your feet? Is it not enough for you to drink clear water? Must you also muddy the rest with your feet?[7]
- The land is mine and you are but aliens and my tenants. Throughout the country that you hold as a possession, you must provide for the redemption of the land.[8]

Concern for endangered species may be traced to a general concern for preserving elements of the wild, including but not limited to threatened and endangered species. Today's concern for protecting and conserving things that are wild can be said to have resulted from a progression through the following stages:

1. *Fear of the wild*: Until recently, people knew little of the wild, despite continuous exposure to the wild and frequent confrontations with it. Large predators attacked and sometimes even ate people. There were (and still are) poisonous snakes, poisonous plants, and other physical threats from the wild. Before advancement of our knowledge of biology and ecology, all of these threatening but poorly understood wild threats were something to be feared. American society generally existed at this stage only in frontier settings. Not only would an Endangered Species Act be inconceivable to a society at this stage, the polar opposite of actually targeting species for extinction would be far more compatible.

2. *Conquering of the wild*: Until the early twentieth century, wild places represented potentially unused assets: Until forests could be cleared, wetlands drained, and rivers dammed, these resources could not be used to produce wood, farmland, navigation, and other economically valuable goods and tools. People seeking to realize the economic potential of these natural resources did not fear the wild places they sought to conquer; the losses of these wild places were simply an unavoidable collateral loss. America was at this stage over most of the nineteenth and first half of the twentieth century. As with the previous stage, an Endangered Species Act would be inconceivable to a society at this stage, although a species would be targeted for extinction only if there was an economic incentive to do so.

3. *Neutrality toward the wild*: Comfortable in an urban or settled, well-established agricultural setting, people had to neither fear nor conquer the wild. They remembered, either themselves or through their parents or other elders, the former need to fear and conquer the wild. But, the mission was accomplished. There was no nostalgia for what once was to be feared or conquered. Wild things simply were not a part of their lives, and they were contented with that situation. America after World War II, until the mid-1960s, was at this stage. Any proposal to protect endangered species would most likely be met with apathy by a society at this stage.

4. *Appreciation of the wild*: Later generations, with less memory of the fear of or need to conquer the wild, however, did have the luxury of nostalgia. They could safely vacation to wild places, enjoy their beauty and solace, and just as quickly return to the safety and abundance of civilization. They returned to civilization and demanded preservation of the wild areas they had visited. The emphasis was purely on preservation, on trying to establish ecological museums. This is where U.S. society was as it enacted the wave of environmental laws of the late 1960s and early 1970s, including the Endangered Species Act. The original bent of the Endangered Species Act reflected this appreciation stage. Threatened and endangered species and critical

habitats were to be left alone, intact and untouched, completely outside the influence of economic activity.

5. *Harnessing of the wild*: Appreciation of wild things is certainly a necessary prerequisite to the Endangered Species Act. And, for conservation idealists, the appreciation stage must seem like the ultimate
achievement. But for conservation to ensure, it must survive in the
real world. Conservation has costs, and someone must bear those
costs. The economic cycles of expansion and recession, prosperity
and depression, inevitably introduce fluctuation into the public's
enthusiasm for conservation.

So, if the initial establishment of the Endangered Species Act was a product of the "appreciation" stage, survival of the act will depend on America's
continued progression into the "harnessing" phase. As explained in Chapter
6, the act as originally implemented had no provisions for incidental take
permits—developers, landowners, and other economic interests were simply
to stay away from endangered species and not interfere with their recovery.

8.3 Basic Sources of Opposition to the Endangered Species Act

Few people are opposed to the concept of protecting endangered species
per se, but the Endangered Species Act has faced significant opposition ever
since its enactment. Opposition to the act, as well as opposition from related
environmental protection states, especially Section 404 of the Clean Water
Act, may be broadly categorized as originating from three sources, one philosophical and two economic:

1. Concern over infringement on private property rights;
2. Concern over depressing economic activity; and
3. Concern over government deficits and debt.

Few opponents are calling for outright repeal of the act. Some may secretly
wish for repeal of the act, but they know repeal would be both unpopular and probably impossible. Many of the arguments for repealing the act
are that regulatory burdens placed on owners of property providing habitat
actually discourage recovery of endangered species because they encourage owners to take whatever actions they legally can to discourage use of
the property by the species.[9] Some opponents make a convincing argument
that private ownership of land harboring endangered species is more likely
to conserve those species than an inattentive government.[10] Many of these

individuals may share the objectives of the act but feel that those objectives can be best met outside a formal government regulatory program.

8.3.1 Private Property Rights

The most cerebral of the three concerns is deeply rooted in American culture. The British North American colonies had been largely settled by people displaced from the then far more crowded England, where landownership had been historically concentrated with a small landed gentry. Although many had to incur significant debt to make the passage, some having to work on other people's land as indentured servants for several years after the crossing, the colonies offered them their only chance to work on their own farms free of having to pay much of the fruits of their labor to a lord or other feudal landowner. On receiving their land, many then had to expend considerable labor to clear it of trees, rocks, and other obstructions to cultivation. Faced with the hardships and isolation of frontier life, the colonists felt little connection to the entrenched government and feudal structure of the motherland.

As the distant government imposed increasing taxes on the colonies, in part to fund protection of the colonies but also to exploit the seemingly limited natural resources of the sparsely populated colonies, the seeds of the American Revolution were born. The author of the Declaration of Independence, Thomas Jefferson, was strongly attached to his land at Monticello and promoted the United States as a community of small-landholding farmers. Most of the other founding fathers were also landowners with strong sentimental ties to their land. America's frontier mentality toward land and landownership was intensified, not diminished, during the nineteenth and early twentieth centuries as the Midwest, West, and ultimately Alaska were settled and land was acquired by individuals from the public domain under the provisions of the Homestead Act (which remained in force as recently as the 1970s). Sentimentality to land and landownership continues to permeate American speech, with expressions originating in American English such as "for land's sake" and "Don't bet the farm on it."

The importance of private property rights is reflected in the Fifth Amendment of the Constitution, which states:

> No person shall be held to answer for a capital, or otherwise infamous crime, unless on a presentment or indictment of a Grand Jury, except in cases arising in the land or naval forces, or in the Militia, when in actual service in time of War or public danger; nor shall any person be subject for the same offence to be twice put in jeopardy of life or limb; nor shall be compelled in any criminal case to be a witness against himself, nor be deprived of life, liberty, or property, without due process of law; nor shall private property be taken for public use, without just compensation.[11]

Because of the wording of the Fifth Amendment, actions that deprive owners of their property without just compensation are termed *takings*. Use of the term *taking* in the context of private property deprivation must not be confused with use of the same term in the context of taking of a species protected under the Endangered Species Act (or under the Bald and Golden Eagle Protection Act and the Migratory Bird Treaty Act).

The Fifth Amendment places deprivation of property on the same level as deprivation of life and liberty. Even in the sparsely populated eighteenth century, the Fifth Amendment recognized that the government would sometimes have to take private property to serve the greater good of the population at large; it does not disparage government acquisition of private property but merely states that such acquisition must be accompanied by appropriate compensation. When the federal or state government acquires privately owned land to build highways, utilities, parks, or other public facilities, it always offers payment to the affected landowner, although establishing the rightful value of the compensation is always a thorny question. The process of acquiring private property to serve public interests is termed *eminent domain*, always requires compensation, and has occurred frequently throughout American history.

However, the Endangered Species Act and most other U.S. environmental statutes rarely force owners of private property to turn over ownership to the government, and if they did, the government would have to offer suitable compensation. But, what they do frequently do is regulate the use of land; they often prevent owners from using their property in a way the property owners desire. Diminution of property value caused by government regulatory actions is sometimes termed "regulatory takings." For most land developers, that is a way that maximizes economic returns, sometimes referred to as the "highest and best use." The issue of whether regulatory actions limiting the ability of property owners to realize the maximal use of their property was first encountered in the early twentieth century in the context of zoning. As cities became increasingly large and crowded, local governments started to establish boundaries to where certain types of development would be allowed (e.g., residential zones, commercial zones, industrial zones, etc.). The constitutionality of zoning, without governments having to compensate landowners for diminished values caused by zoning, was generally settled by a Supreme Court decision in 1926: *Village of Euclid v. Ambler Realty Co.*[12] The case is summarized as follows:

> A suit to enjoin the enforcement of a zoning ordinance with respect to the plaintiff's land need not be preceded by any application on his part for a building permit, or for relief under the ordinance from the board which administers it, where the gravamen of the bill is that the ordinance, of its own force, operates unconstitutionally to reduce the value of the land and destroy its marketability, and the attack is not against specific provisions, but against the ordinance in its entirety.[13]

Although some property owners still regarded zoning and other land use regulation as an infringement on constitutionally protected property rights, the issue was largely dormant until the 1970s and 1980s, with implementation of the Endangered Species Act and, especially, extension of Section 404 of the Clean Water Act to wetlands. Section 404 of the Clean Water Act established in 1972 the ability of the U.S. Army Corps of Engineers (USACE) to regulate "waters of the United States" extending beyond the traditional navigable waters covered under the Rivers and Harbors Act of 1899, but the question of whether such waters included adjacent wetlands was not resolved until 1985 with *United States v. Riverside Bayview Homes, Inc.*[14] Attempts by USACE to establish a technically defensible process for delineating the landward extent of adjacent wetlands culminated with the publication in 1987 of the *Corps of Engineers Wetlands Delineation Manual* (the 1987 manual).[15] Wetland delineation using the 1987 manual and some earlier procedures led to the identification of many seasonally dry areas as wetlands, thereby requiring developers to obtain permits for activities proposed for privately owned land areas such as swamps and marshes not traditionally considered to be bodies of water.

Furor among development and private property rights interests intensified following publication in 1989 of an interagency wetland delineation manual[16] intended to satisfy the wetland delineation needs not only of USACE but also the Fish and Wildlife Service (FWS), which was developing the National Wetland Inventory at the time; the U.S. Environmental Protection Agency (EPA), which has oversight and "veto" authority over USACE in its administration of Section 404; and the U.S. Department of Agriculture Soil Conservation Service,[17] which administered wetland provisions under the Food Security Act of 1985.[18] While the intent of the 1989 manual was not to increase the scope of lands over which Section 404 would apply, clarifications to the three-parameter procedure from the 1987 manual did lead some wetland delineators to establish wetland boundaries over substantially drier lands than would have previously been included using the earlier manual. At least part of this expansion of wetland coverage was likely attributable to a misunderstanding by inexperienced wetland delineators of the more technical language used in the 1989 manual. I performed several wetland delineations using both manuals and generally feel that the wetland boundaries identified in most settings by either manual are substantially identical, as long as both manuals are properly followed by qualified biologists.

The response of business and property rights interests to the 1989 manual did not go unnoticed by politicians, including President George H. W. Bush and, in particular, Vice President Dan Quayle. The vice president headed up a Competitiveness Council established to relieve businesses and landowners of environmental regulatory burdens. Regarding wetlands, Quayle stated:

> Another initiative at the Competitiveness Council was to oversee the rewriting of the Federal Wetlands Manual [1987 manual]. President Bush had promised "no net loss" of wetlands, and people in [environmental]

agencies tried to widen the definition of wetland. This caused thousands of acres of dry land to be reclassified as wetland, including a lot of perfectly tillable Indiana farmland. We asked the bureaucracy to consider a revolutionary idea: if the land isn't wet, maybe we shouldn't call it a wetland.[19]

Although best known for his positions on wetland regulation, Vice President Quayle expressed concern over similar effects from the Endangered Species Act as well. He stated:

The Endangered Species Act is another example of using bad means to achieve good objectives. If you have endangered species on your property, chances are that the government will come in and stop you from using your property. Some landowners actually cut down trees and bushes on their property rather than risk making it hospitable to endangered species. That is wrong and indefensible. Citizens who put forth the effort to maintain habitats ought to be rewarded, not threatened with financial ruin.[20]

The habitat conservation plan process for obtaining incidental take permits had already been in place for several years prior to this statement, hence theoretically the government could not actually stop development of land containing endangered species unless it was impossible to prevent actual jeopardy to the recovery of the species. But, the costs of mitigation measures required to support a habitat conservation plan acceptable to the Services could be so onerous that it made development economically impracticable. The safe harbor provisions and no surprises rule of the late 1990s did not yet exist, but Quayle's statement clearly presaged these future improvements to the act.

8.3.2 Concern over Depressing Economic Activity

The discussion of property rights in Section 8.2.1 is primarily directed to the ethical issues of protecting property rights, as protected under the U.S. Constitution and engrained into American culture. An interrelated issue is how enforcement of the Endangered Species Act can dampen economic activity by imposing restrictions on land uses. Most landownership is an investment; regulatory actions that prevent landowners from achieving the highest and best use of their property diminish the economic return from that investment. The concern, while primarily directed at restrictions on use of private land, can also extend to restrictions on public land, especially public land managed by the U.S. Forest Service and Bureau of Land Management.

Everything discussed previously is therefore economic as well as philosophical in character. However, the economic repercussions can extend far beyond those who own the affected property—they can extend to regional economies as well, depressing employment and the service industries that support that employment. The employment and regional economic effects of environmental regulation can perhaps be best illustrated by the northern spotted owl (*Strix occidentalis caurina*) controversy of the early 1990s, which was coincident with the wetlands controversy of that same time described in Section 8.2.1.

The northern spotted owl ranges over much of the western forested mountains of Washington, Oregon, and northern California as well as parts of the Canadian province of British Columbia. Its favored habitat is older, established forest cover with moderate-to-high canopy closure, large overstory trees with various deformities such as large cavities, and open space beneath the canopy to allow flight.[21] Its habitat requirements are therefore in direct conflict with harvesting economically valuable old-growth timber in a region where logging remains an important source of jobs and a key player in rural economies. The primary threat to the species has long been considered to be loss of the mature forest cover, as results from timber harvest but can also result from other causes such as intensive wild fires, but researchers are also recognizing that increased completion with the barred owl (*Aix stricta*) may also be contributing to population declines.[22] The barred owl is a species of forests of the eastern United States that has been expanding its range to the Pacific Coast in recent decades. While there were 5.431 known sites inhabited by the northern spotted owl as of July 1994, only 1070 sites were recognized as of June 2004.[23] While uncertainty remains over the causes of the decline, whether attributable to logging, the barred owl, or a combination of these and possibly other unknown causes, the species is clearly declining.

The FWS announced its proposal to list the northern spotted owl as threatened on June 23, 1989.[24] Numerous commenters on the proposed listing expressed concern that listing the northern spotted owl would impose economic hardships on communities that benefit directly and indirectly from harvesting old-growth timber. The FWS responded that a listing decision must be "based solely on biological criteria and to prevent non-biological considerations from affecting such decisions" and that "economic considerations have no relevance to determinations regarding the status of species."[25] While development and economic interests had been suspicious of the Endangered Species Act from the start, the northern spotted owl listing clearly cast the antagonisms between the act and economic interests in the public light. The FWS succeeded in listing the species as threatened, but the controversy of jobs over a species became elevated to the national stage, with frequent newspaper articles and news coverage that galvanized populist opposition to environmental regulation in a way that only the wetlands issue had in the prior years.

The northern spotted owl issue primarily involves the logging and the timber industry against the Endangered Species Act, but the controversy over another species pits the act against land development and agricultural interests and overlaps even more directly with the wetlands controversy: the California red-legged frog (*Rana aurora draytonii*). This amphibian is a species of wetland

and other aquatic habitats endemic to California and northern parts of the Mexican state of Baja California. Its breeding sites include "pools and backwaters within streams and creeks, ponds, marshes, springs, sag ponds, dune ponds and lagoons" as well as "artificial impoundments such as stock ponds."[26] Sentiment for protecting the frog against extinction extends beyond the usual scientific arguments and into the cultural realm as well. The California red-legged frog is generally considered to be the frog in Mark Twain's classic short story, "The Celebrated Jumping Frog of Calaveras County."[27]

The main controversy over the California red-legged frog is not its listing but proposals to designate critical habitat in California's central valley that would limit the extent of both agricultural and suburban land development.

8.4 Specific Recent Controversies

I have been completing field studies and authoring technical reports on ecological resources, including wetlands and endangered species, since the late 1980s. Effective practice as an environmental consultant requires keeping up with emerging controversies and proposals to change the environmental statutes and regulations that drive his or her profession. The following section discusses a few of the most visible controversies I experienced over the span of my professional career. It is not intended to be a comprehensive list. However, the controversies included provide an informative cross section of the controversies that have shaped current practice under the Endangered Species Act and related environmental statutes.

8.4.1 Republican Contract with America

The conflicts between private property and business interests and regulatory takings under the Endangered Species Act and Section 404 of the Clean Water Act reached a crescendo in the national media during the 1994 midterm congressional elections. The Contract with America,[28] the broad statement of policy objectives targeting smaller, leaner government that many Republican congressional candidates used to win election to and gain control of the House of Representatives in the 1994 elections, does not specifically mention the Endangered Species Act or related environmental statutes such as the National Environmental Policy Act (NEPA) or Clean Water Act. Contrary to what some thought at the time, the contract and its package of proposed bills did not call for outright repeal of these or any other environmental protection statutes. However, one component bill included within the contract, the Job Creation and Wage Enhancement Act,[29] contained two provisions that if enacted could have substantially altered how the Endangered Species Act was administered. Once the Republican 104th Congress was in place, the Congress forwarded the private property compensation principles of the Job Creation and Wage Enhancement Act by proposing the Private Property Protection Act of 1995, which stated:

> The Federal Government shall compensate an owner of property whose use of any portion of that property has been limited by an agency action, under a specified regulatory law, that diminishes the fair market value of that portion by 20 percent or more. The amount of the compensation shall equal the diminution in value that resulted from the agency action. If the diminution in value of a portion of that property is greater than 50 percent, at the option of the owner, the Federal Government shall buy that portion of the property for its fair market value.[30]

The original version of the bill proposing the Private Property Protection Act of 1995 required compensation for agency actions that reduced the value of property by 10% or more.[31] The final bill sent by the House to the Senate set the bar for compensation to a 20% diminution. Either way, this requirement would have substantially increased the costs for imposing limitations on property use or mitigation requirements on private property owners. Considering that the Republicans were simultaneously seeking to substantially reduce the budgets of most federal agencies, the net effect of the property compensation requirement would almost surely have been to inhibit agencies such as FWS from imposing substantial limitations or mitigation on permit applicants. The measure would not have repealed the Endangered Species Act, but it would still have substantially weakened it, and without having to actually amend it. Congressional representatives would not have had to openly oppose the Endangered Species Act or other environmental protections favored by substantial blocks of their electorate. The effect on Section 404 would have likely been even greater than for the Endangered Species Act; the USACE would have had to pay out whenever it denied a permit application to build in privately owned wetlands or imposed significant wetland mitigation requirements on private developers.

The Private Property Protection Act of 1995 was never enacted. In fact, the Contract with America ultimately had little direct effect on the Endangered Species Act or related environmental statutes. However, two key new provisions were enacted into the Endangered Species Act in the second term of the Clinton administration, the no surprises rule and the safe harbor rule, both described in detail in Chapter 6 of this book. Both of these provisions were commonsense modifications to the Endangered Species Act, making it more workable in the real world and blunting some demands for repeal. The ultimate legacy of the Contract with America may have been to save the Endangered Species Act by making it more practical and less idealistic.

The ultimate effects of the Private Property Protection Act of 1995 on the Endangered Species Act, had it been implemented in the form delivered by the House to the Senate, are hard to estimate. Considering federal budgetary constraints, it may have made mitigation in the form of setting aside land for purely conservation purposes effectively impossible. Some development interests may have still been willing to volunteer areas of their property for preservation to avoid extended paperwork and negotiations with the Services; the new act would not have prohibited voluntary donation of private property value. Relatively inexpensive mitigation measures such as planting tree cover, establishing riparian forest buffers, avoiding work in nest areas during specific months, transplanting or relocating individuals, or establishing nest boxes would likely have remained viable and practicable since they may not have decreased property values by more than 20%. Had Congress and the courts allowed for establishment of a 20% ceiling, relative to the overall value of the property to be occupied by a project, on the costs of natural resource permitting and mitigation, the new act might have been a workable, commonsense compromise. Developers and business owners generally appreciate economic predictability; the costs of environmental regulation and mitigation are less objectionable if they are predictable and can be knowledgably factored into project budgeting. However, had the 20% reduction in value threshold been applied to only that subset of a property containing wetlands, then it would likely have effectively eliminated any meaningful environmental regulatory protections.

The ultimate effects of the 1994 elections and the Contract with America on the Endangered Species Act are also difficult to assess. None of the radical proposals to weaken the act or related environmental statutes such as the Clean Water Act succeeded. Perhaps the most lasting legacy of 1994 is that the Endangered Species Act and the overall concept of environmental conservation became widely ingrained in the public mind as a partisan issue, supported by the Democratic Party and opposed by the Republicans. Previously, environmental conservation had not been a highly partisan issue and had vocal supporters and opponents in both parties. Congressman John Dingle, the original author of the Endangered Species Act, was a Democrat, but the act was signed into law with enthusiastic support by President Richard Nixon, a Republican. If there had been a significant divide between supporters and opponents of the act prior to 1994, it was urban versus rural, not Democrat versus Republican. Of course, the urban and rural divide persists, but since 1994 there has been a substantial partisan divide as well.

8.4.2 Solid Waste Agency of Northern Cook County, Rapanos, and Other Limitations on Section 404 Scope

Although not directly involving the Endangered Species Act, a Supreme Court decision in January 2001 substantially reduced the scope of Section 404 of the Clean Water Act and in the process dramatically altered the entire environmental planning process in the United States. The 5-4 decision[32] was highly divisive between conservation and property rights activists, being separated in timing by only about a month since the even more divisive Bush versus Gore 5-4 decision that ended the prolonged 2000 presidential campaign that had already alienated the Democratic base that includes a large proportion of those who supported aggressive environmental regulation.

Throughout the late 1980s and the 1990s, the USACE had exerted Section 404 jurisdiction over nearly all wetlands meeting the technical criteria for delineation as wetlands. However, the Clean Water Act statutory language limits the act's jurisdiction to "waters of the United States" presently or formerly used for navigation or otherwise involved in interstate commerce. The Riverside Bayview Homes decision[33] and other case law in the late 1980s had established that wetlands "adjacent" to navigable waters could be regulated under Section 404. Wetlands at the edge of tidal waters, rivers, streams, and navigable lakes were clearly adjacent for Section 404 purposes. However, many wetlands in the United States occupy low areas lacking any surface flow connections to navigable waters, including many high-value wetlands with respect to wildlife habitat such as prairie potholes in the Midwest, cypress domes and Carolina bays in the Southeast, and playa lakes in Texas and the Great Basin area of the Southwest. Many of the most popularly valued endangered species, such as the whooping crane (*Grus americana*) and wood stork, depend on such wetlands.

To claim that such wetlands were adjacent and that their protection was necessary to protect interstate commerce, USACE and EPA had used the "migratory bird rule," an argument that these wetlands were used as habitat and for breeding by migratory birds that were (or could be) traded in interstate commerce.[34] Specifically, the migratory bird rule asserted that wetlands could be brought under Section 404 jurisdiction if they could be used as habitat by birds protected by the Migratory Bird Treaty Act, could be used as habitat for endangered species, or could be used to irrigate crops sold in interstate commerce. If the legal basis of the migratory bird rule was tenuous, the scientific justification was sound: The behavior of species is irrespective of state (and even country) boundaries, and failure of even one state to protect habitat for some species could lead to extinction despite the effectiveness of conservation measures in other states.

A challenge to the migratory bird rule was inevitable. It came when the Solid Waste Agency of Northern Cook County (SWANCC), which provides landfill space to part of the Chicago metropolitan area, applied to deposit landfill waste in a portion of its excavated wetlands that had pooled runoff and formed wetlands. SWANCC argued that it should not have to get a Section 404 permit or perform the requisite mitigation because the affected wetlands were not adjacent to navigable waters. Their case went before the Supreme Court. In *SWANCC v. Army Corps of Engineers*,[35] the majority of the Court agreed that the migratory bird rule was not adequate justification for extending Clean Water Act jurisdiction to the subject wetlands. In the majority opinion, Justice Rehnquist stated that "permitting respondents to claim federal jurisdiction over ponds and mudflats falling within the 'Migratory

Bird Rule' would result in a significant impingement of the States' traditional and primary power over land and water use."[36]

The decision struck down use of the migratory bird rule as a basis for adjacency under Section 404, and strictly speaking that was the substantive extent of the ruling. But, its effects went much further; it emboldened property rights activists to take ever-stronger stands against Section 404. The forum on the Society of Wetland Scientists Web site[37] exploded with vigorous debates by scientists both for and against the SWANCC decision, overwhelming the traditional use of the forum to discuss more technical issues of wetland delineation, mitigation, and research to the extent that the society ultimately had to limit posting access. SWANCC eclipsed the efforts in the early 1990s to restrict Section 404 by reconstituting the wetland delineation process; the value of limiting the spatial extent of wetland delineation was largely moot if the entire wetland could be excluded from regulation. Several years of uncertainty ensued following the SWANCC decision in which USACE would attempt to exert jurisdiction over wetlands isolated from navigable waters and their stream systems by arguing various bases other than the migratory bird rule, such as proximity to navigable waters or connection via the groundwater. Ultimately, USACE converged on asserting jurisdiction only over those wetlands from which surface flow could follow an unbroken pathway, not blocked by uplands, to a navigable water body, regardless of how long that pathway might be. They also continued to assert jurisdiction over some wetlands lacking a direct surface connection if they were positioned in a floodplain or were separated from adjacent wetlands only by artificial blockages such as levees or elevated roadbeds.

Just as the uncertainty over SWANCC had begun to be resolved, the Supreme Court issued an even more far-reaching decision in *Rapanos v. United States.*[38] Rapanos was a developer in Michigan who argued that he should not have to get a Section 404 permit to fill a series of ditches containing wetlands even though the ditches ultimately flowed into navigable waterways more than 20 miles away. Rapanos argued that the wetlands in the ditches should not be considered adjacent to the navigable waters because of the long distance and tenuous physical relationship. The Supreme Court responded with an unusual 4-1-4 "plurality" decision in which the opinions of four justices ruled with Rapanos, four dissented, and one issued a separate ruling. Wetland scientists and other environmental consultants who may have hitherto considered the names of the Supreme Court justices to be a trivia question suddenly attached them to key positions regarding Section 404 jurisdiction. The majority opinion, authored by Justice Scalia, basically stated that wetlands had to directly border navigable waters to be regulated under Section 404. Such wetlands are sometimes termed "Scalia" wetlands. Wetlands adjacent to nonnavigable reaches of streams, such as the wetlands on Rapanos's property, were nonjurisdictional.

However, Justice Kennedy disagreed with Scalia on the limitation of adjacency to wetlands directly bordering navigable waters. He stated that wetlands bordering nonnavigable tributaries to navigable waters could fall under Section 404 jurisdiction if they had a "significant nexus" to interstate commerce, that is, if their degradation could substantially affect the navigable waters used in interstate commerce. Determining whether a significant nexus occurred involved completing a lengthy form, sometimes termed a "Rapanos form," involving a battery of questions regarding hydrological position in the landscape and physical and biological characteristics of the wetlands and associated stream reach.

Regardless of one's position regarding the desirability of maintaining or restricting the scope of Section 404, the SWANCC and Rapanos decisions had the unfortunate result of focusing too much attention on the arbitrary position of wetlands with respect to navigable waters rather than on whether protection of the wetlands had scientific merit. The relative value of individual wetlands, whether with respect to endangered species, flood control, or other physical and biological functions, was immaterial to the argument; everything revolved around how far from a navigable water a wetland could be and still receive protection under Section 404. Many environmental consultants, myself included, found that completing Rapanos forms[39] for delineated wetlands could take as much billable time as completing the traditional wetland delineation forms. Efforts directed to answer this somewhat-arbitrary question constituted resources that could have been better spent on designing projects to avoid wetlands and on mitigation.

The effects of the SWANCC and Rapanos decisions on endangered species have yet to be played out. Clearly, they have reduced protections for large areas of wetlands providing valuable habitat for several endangered species, such as the whooping crane, bog turtle, and wood stork. This is especially true in many southern and western states that lack state-level wetland protection statutes unlinked to the federal Clean Water Act. Perhaps more important, the SWANCC and Rapanos decisions have emboldened many developers and property owners to resist environmental regulatory protections in court rather than working through the regulatory system to obtain permits and complete requisite mitigation. Money spent on court cases is money that might have otherwise been spent more productively on mitigation benefiting endangered species and other ecological resources.

8.4.3 Proposed Threatened and Endangered Species Recovery Act of 2005

Representative Richard Pombo of California, a ranch owner and property rights activist, sponsored HR 3824, a bill to "amend and reauthorize the Endangered Species Act of 1973 to provide greater results conserving and recovering listed species."[40] The bill would have reduced the regulatory burden on private property owners and action agencies, raised the bar for listing new species, shifted more of the cost of complying with the act from private property owners to the government, and established clearer targets for achieving recovery of species that are listed. The bill was clearly written from the perspective of private property owners faced with expenses and uncertainties associated with the act. It followed the familiar pattern of reducing the regulatory and bureaucratic burden of the Endangered Species Act while accomplishing its overall objective of recovery of species in imminent danger of extinction.

The most far-reaching and controversial provisions of the bill were the following:

- It would have tightened criteria for listing new species by requiring an analysis "of the economic, national security, and other relevant impacts and benefits of" listing. This change would have introduced economic and national security considerations into listing decisions traditionally driven only by survival and recovery of species. This provision would have likely prevented listing of some species otherwise in danger of extinction because of the possible costs or national security implications of extending the act's protections to those species. New listings may have been largely limited to declining species only found on public conservation lands or in geographically remote areas with little agricultural, forestry, resource extraction, or development potential. The actual effects of this provision may not have been far reaching, as new species listings have from the outset of the act unofficially been subject to the politics of costs and economics.

- It would have repealed "the authority of the Secretary to designate critical habitats." It is unclear whether this provision would have removed regulatory protections for already-designated critical habitats or just prevented the designation of new critical habitats. Either way, it would have reduced the protections under the act targeting habitats rather than individual species.

- It would have required that the "terms and conditions" (i.e., mitigation) required by incidental take permits, whether obtained through the Section 7 consultation process or through the habitat conservation plan process, to be proportional to the specific take authorized under the permit. This was an attempt to place bounds on what can almost be a "blank check" for the Services to extract mitigation measures from a few developers for species whose decline is resulting from uncharacterized cumulative effects. Although one might expect private property activists to not view the Section 7 process as controversial, recall that private developers requiring federal permits, such as Section 404 permits, can be drawn into the Section 7 process and be bound by the terms and conditions of incidental take permits issued at the conclusion of the Section 7 process.

- It would have authorized the secretary of the interior "to provide grants to promote the voluntary conservation of endangered or threatened species by private property owners" but also required the secretary "to compensate such owners for the cost to them of conservation measures imposed by this Act."
- It would have required the secretary of the interior to "reimburse owners of livestock for any loss of livestock resulting from reintroduction of endangered and threatened species into the wild." This provision was primarily driven by concerns that reintroduction of the gray wolf (*Canis lupus*) to public lands would lead to predation on livestock held on nearby private lands. Gray wolves are apex predatory carnivores that travel long distances in search of prey and that cannot distinguish public from private property or livestock from wild animals.

The bill also included several provisions to streamline bureaucratic procedures under the act and increase the transparency of the act, especially with respect to listing decisions that could potentially affect private property owners. Although passed by the House of Representatives, the bill was never approved by the Senate or sent to the president for signature. The bill's sponsor, Richard Pombo, was defeated in the 2006 midterm congressional elections, and control of the House passed from the Republicans to the Democrats. The bill did not therefore alter the Endangered Species Act or its regulations.

8.4.4 Polar Bear Listing

Perhaps no listing action for a single species under the Endangered Species Act has engendered the level of controversy as the decision by FWS in 2008 to list the polar bear (*Ursus maritimus*) as threatened.[41] This listing action thrust the Endangered Species Act, long controversial for its effects on land use and property rights, into the center of the very different and even more divisive controversy over global warming and greenhouse gas emissions. The polar bear listing was the first time that a species was listed under the act solely on the basis of speculative claims using computer models regarding effects on the species resulting from possible future climate change.[42] The controversy was intensified in 2010 when FWS designated approximately 187,157 square miles of Alaska and associated territorial waters as critical habitat for the polar bear.[43]

FWS issued separate regulations under the Endangered Species Act for the polar bear; basically, the special rules exempt incidental take activities from the usual regulatory protections under the act as well as allow continued hunting if in accordance with other federal laws and international treaties.[44] The regulatory protections established for the polar bear under the act are quite limited and, by themselves, would not be expected to be especially controversial. However, listing the polar bear can be viewed as one of several incremental steps to tightening federal control over business interests in the name of climate change. Many fear that U.S. business interests will be significantly hobbled by new regulations limiting carbon dioxide emissions, thought by many scientists to be causing noticeable climate change, formerly referred to as "global warming." The issue has been accompanied by a lot of hyperbole, such as images of flooded coastal cities and agricultural lands transformed into deserts. A book on the Endangered Species Act is not the place for an in-depth analysis of this highly complex issue, except to note that the act, like so much else in American life today, is becoming increasingly involved in this larger, more global controversy.

The Endangered Species Act is likely to continue to be increasingly intertwined with the climate change issue in the near future. If climate conditions do shift northward as many of the controversial computer modeling studies suggest, some species could have trouble adapting to rapidly changing geographic ranges and to changing environmental conditions in existing ranges. Historical patterns of nesting and migration could be rapidly altered for some species. Populations of some species could decline, leading to increased petitioning for listing of ever more species under the act. Even if the climate changes do not occur, or do not occur as rapidly as feared, the perception of a threat from climate change on many species may still lead to increasing rates of new listing petitions.

8.4.5 Bush Administration 2008 "Midnight Rule Changes"

In its closing months, the Bush administration surprised many working with the Endangered Species Act by issuing a series of modest regulatory changes to the interagency consultation requirements under Section 7. It is not unusual for outgoing presidential administrations to issue a series of last-gasp executive branch rulings that favor the interests of that administration. These rule changes are sometimes termed "midnight rules." The outgoing Bush administration was more interested in reducing the regulatory burden on business than the incoming Obama administration. One motivating factor for these changes was the earlier listing of the polar bear, discussed previously. If the increased regulatory burdens anticipated to address climate change were at least partially ensconced into the Endangered Species Act compliance process, then one way to reduce the new burdens would be to simplify the Endangered Species Act process itself.

The midnight rule changes would have substantially changed how environmental consultants and agency staff biologists work with the Endangered Species Act. Although the rule changes were enacted by the outgoing Bush administration, they were withdrawn by the Obama administration very shortly after inauguration. The changes had not been in place long enough to have substantially altered established practice trends related to the Section 7 consultation process. The more substantive changes included the following:

- Action agencies would not be required to perform formal consultation if their own analysis indicated that the direct or indirect effects of the proposed action would not result in take of listed species or critical habitats;

- Action agencies would not be required to perform formal consultation for proposed actions whose effects were "manifested through global processes" and were minimal; and

- The definition of a biological assessment used in the Section 7 process would be changed to allow use of other documents such as environmental impact statements (EISs) or environmental assessments (EAs) that contain the relevant information.

The original author of the Endangered Species Act, Rep. John Dingell of Michigan, who has remained in Congress continuously since passage of the act and who has been one of the most vocal congressional promoters of environmental protection, weighed in on the midnight rule changes. He stated:

> I applaud President Obama's action today to restore the integrity of the Endangered Species Act. When I wrote this legislation in 1973 and it was passed into law, it was one of the proudest moments of my career. When the Bush Administration, in the final days of their White House tenure, attempted to dismantle the law, I was dismayed and called on the Bush Administration to halt their damaging rulemaking which stripped the law of an important requirement for federal agencies to consult with scientific experts on projects they undertake. Today's actions by President Obama put us on the right track to a proper policy that will protect our natural heritage while also preserving our economy—a delicate balance that I tried hard to strike in the bill. It has worked; more than two dozen species of plant and wildlife have been saved by the ESA, including the Bald Eagle and the Gray Wolf, making our world a better and richer place to inhabit. And we have done so without jeopardizing jobs or our economy. I stand ready to work with President Obama and Interior Secretary Ken Salazar to ensure that the ESA is a strong law that all Americans can be proud of in the future.[45]

Although the changes would not have dismantled the act, they would have weakened the involvement of scientific experts in many decisions made potentially affecting endangered species. They would not have "stripped" the requirement for consultation with experts, just limited the situations in which consultation would be required. Mr. Dingell's statement reflects a view that the Endangered Species Act is a carefully crafted statute that has worked well and proven itself over more than three decades. The original author of the act seems pleased with the development and maturation of the act and does not see a need for hasty tinkering. Considering that Mr. Dingell represents a district in metropolitan Detroit that depends heavily on manufacturing, his statement that the act does not conflict with the economy carries some weight.

Despite their ominous-sounding name and strong derision by environmental activists, certain elements of the rule changes might have helped further as well as just hinder the objectives of the Endangered Species Act. The proposed regulatory changes were not necessarily all detrimental to the objectives of the act, and some might have helped deflect some of the intensifying criticism of the Endangered Species Act that came about in the years following the 2008 election. Each of the three substantive proposed changes is discussed next.

8.4.5.1 Reduction in Formal Section 7 Consultation Requirements

As described in Chapter 5 of this book, the Services may draw three possible conclusions regarding possible adverse effects on listed species or critical habitat: no effect; may affect, not likely to adversely effect (NLAA), or may adversely affect (MAA). If, after review of the best-available relevant scientific information, an action agency believes that the Services would conclude no effect for all potentially affected species and habitats, then the action agency has traditionally not been required to do formal consultation. Because the action agency is still held responsible for avoiding unpermitted take of listed resources, most have voluntarily sought concurrence from the Services even for actions they consider to be no effect.

For situations where the expected Services conclusion was NLAA or MAA, action agencies have traditionally initiated formal consultation. The midnight rule changes would have changed this paradigm by allowing action agencies not to formally consult if the best-available science indicated that the appropriate conclusion was NLAA (they traditionally did not consult if the expected conclusion was no effect). Agencies would still have had to consult formally if the best-available science indicated a conclusion of MAA.

Proponents of the rule changes argued that they would reduce the regulatory burden on agencies proposing actions with minimal or no potential for adverse effects on listed species or habitats while also allowing the Services to focus their efforts on those actions with the greatest potential for adverse effects. Critics argued that the rule changes would increase the potential for abuse by agencies promoting development activities, such as oil and gas drilling. The criticism was harsh; one critic, the World Wildlife Fund, claimed that the rule changes "would eliminate a key environmental review process that ensures federal development projects do not cause additional harm to species that are at risk of extinction."[46] However, the rule changes would not have removed the responsibility of action agencies for ensuring compliance with the act's prohibitions on take or on jeopardizing the recovery of listed species. Agencies lacking the requisite expertise would therefore likely have been motivated either to seek qualified consulting services or to voluntarily engage the service's experts through informal consultation.

I had just begun employment as a staff biologist with a federal agency (the Nuclear Regulatory Commission) about six months prior to announcement of the rule changes. I had worked for about 20 years previously in the environmental consulting arena, gaining a lot of knowledge and insight preparing reports for use by agency officials but never possessing any regulatory authority. I had been excited that the new rules would have thrust greater authority under Section 7 on action agency staff biologists such as myself, pulling a portion of that authority away from the Services. The changes may therefore have worked well for agencies with a diverse and highly experienced staff of internal biologists, all with their professional reputations to defend, but not with some smaller agencies. Furthermore, the effect might have been for even some large agencies to delegate the responsibilities to general managers rather than qualified biologists.

8.4.5.2 No Formal Consultation for Effects Manifested through Global Processes

The element of the rule changes regarding no formal consultation for effects manifested through global processes was directed squarely at the polar bear, which had just been listed as threatened. Some business interests were worried that the Services might use the Section 7 consultation process to extract concessions related to reducing carbon emissions thought to be adversely affecting polar bears. This change would have also helped alleviate some of the concerns of politically connected oil and gas drilling interests working in areas of Alaska where the polar bear ranges. This attempt to decouple the Endangered Species Act from the climate change controversy might have helped to spare the act some of the animosity received since 2008 from anti-climate change activists.

8.4.5.3 Allowing EISs and EAs to Serve as Biological Assessments

Many environmental professionals experienced in preparing EISs or EAs and accompanying biological assessments agree that there is usually a lot of duplication in the text of the two documents. A strong argument can be made for streamlining biological assessments by allowing them to incorporate by reference relevant text from NEPA documents addressing the same action. Such streamlining is consistent with goals promoted by the Council on Environmental Quality NEPA regulations[47] in the 1970s. Simply providing an EIS or EA and stating that it is—as is—a biological assessment may not work effectively in most instances. Biological assessments require a lot of information not always included in EISs or EAs. EISs and EAs are documents written for the general public that summarize potential impacts from alternatives; biological assessments are technical information documents written for experts with the Services and focus only on the proposed action, not alternatives. If an EIS also had to serve as a biological assessment, it would have to be substantially expanded to include the technical information required by the Services to support their consultation.

This expansion would run counter to ongoing attempts to make EISs more concise and in plainer language to communicate better with the public. Stretching an EA to simultaneously serve as a biological assessment would be even more problematic; many EAs use less than 30 pages to address all environmental resources, not just biological resources. The result would likely be more "super EAs," EAs of considerable length that many NEPA practitioners would like to eliminate.

8.4.6 The Tea Party Movement and 2010 Pledge to America

The smaller government theme promoted by the Republican Party in the 1994 midterm congressional elections with the Contract with America was repeated by the Republicans in the 2010 midterm congressional elections with the Pledge to America. The Pledge to America was largely formulated by a fiscally conservative but generally libertarian wing of the Republican Party that had coalesced as the Tea Party after the election of Barack Obama. The Tea Party is not a political party but a group of Republicans promoting values such as limiting the scope and power of the federal government, balancing the national budget, and paying down the national debt. The Tea Party does not advocate outright repeal of key environmental protection statutes. Like the Contract with America, the Pledge to America does not mention the Endangered Species Act or other environmental statutes by name. In fact, while compensation for regulatory diminishment of private property value was addressed closely by the contract, the pledge does not directly address private property compensation (although most of its backers most assuredly would favor statutory requirements for compensation for property values reduced due to regulation.)

However, reducing federal regulations is a key objective of the Tea Party. With respect to environmental regulations, the imposition of regulations restricting carbon emissions (commonly termed "cap and trade") is at the forefront of the Tea Party's resistance. However, with the listing of the polar bear, the Endangered Species Act became intertwined with the global climate change debate and hence with the Tea Party's principal environmental target. But, even without any association with global climate change, the traditional regulation of private property use associated with the Endangered Species Act is still a Tea Party target, if not its most visible. The Tea Party can be thought of as Contract with America 2.0, with similar environmental objectives. The Tea Party is just the current manifestation of opposition to environmental regulations by property rights activists and business interests. The swinging pendulum between environmental and property rights activists is unlikely ever to disappear as long as environmental regulations exist.

8.4.7 The Endangered Species Act and the 2012 Presidential Election

By the time this book is published, the 2012 elections will be over or at least close to over. The timeliness of a detailed discussion of the role of the Endangered Species Act in the election will be fading. However, the election still illustrates much about the likely future of the act and its associated controversies in the next several years. An in-depth discussion of how the act is perceived by each of the major candidates will therefore prove useful for many years after conclusion of the elections.

As of the writing of this book, the 2012 presidential and congressional elections are playing out as Tea Party interests vying against traditional Republicans and Democrats seeking to continue the programs of the Obama administration. Environmental issues are not at the forefront, except for the highly contentious issue of global climate change and proposals to regulate carbon emissions. However, President Obama's decision to delay the proposed Keystone oil pipeline to carry oil from the Canadian oil sands to Texas oil refineries until NEPA issues are resolved has reelevated NEPA to the forefront as well as increased scrutiny of how environmental regulations are affecting domestic energy production.

8.4.7.1 Mitt Romney

The Web site for Mitt Romney's campaign does not specifically address the Endangered Species Act but speaks extensively of the costs of regulatory burdens on the economy. Regarding environmental regulations, it states:

> As president, Mitt Romney will eliminate the regulations promulgated in pursuit of the Obama administration's costly and ineffective anti-carbon agenda. Romney will also press Congress to reform our environmental laws to ensure that they allow for a proper assessment of their costs.[48]

Any interest that Romney would take in the Endangered Species Act would probably be primarily from the perspective of reducing its effects on the economy. He does not have a history of vocal opposition to regulatory taking from a philosophical perspective. But, as a businessman faced with intense public concern over employment and the economy, Romney can be expected to scrutinize all environmental regulations, including but not limited to the Endangered Species Act, for opportunities to reduce economic burdens on small and large businesses, including businesses involving landownership. Romney can also be expected to seek opportunities to alleviate burdens to the struggling real estate industry, whose travails have underlain so much of the post-2008 recession.

8.4.7.2 Ron Paul

The most aggressively libertarian of the Republican candidates, Ron Paul's Web site, like those of the other candidates, does not specifically address the Endangered Species Act. It does, however, call for the elimination of the EPA and states that, "Polluters should answer directly to property owners in court for the damages they create—not to Washington."[49] This suggests that Ron Paul would likely seek to eliminate or simplify the paperwork and mitigation burden for interests seeking incidental take permits for private property, and he would likely seek to decouple private-sector permit applicants from the Section 7 consultation process. He would likely reduce the applicability of Section 404 to private-sector developers, indirectly reducing their exposure to Section 7 even if the Section 7 process itself is left unchanged. There is little in Dr. Paul's platform, however, that suggests that the basic framework of the Endangered Species Act would be altered.

Ron Paul has said little on the Endangered Species Act, but a statement he made during the 2008 presidential elections reveals much about how he views the act and what he would at least desire should he be elected president. At a speech in Seattle in September 2007, while Ron Paul was a candidate in the 2008 presidential election, he stated:

> I've been reading the Constitution now and then. I can't find endangered species written in the Constitution and I don't think that's a federal function. But that doesn't mean that if we're not for the Endangered Species Act we shouldn't be interested in protecting species. We should be doing it in a private sort of way. Sometimes … if there's an endangered species you say "Well, I'm going to raise a few of those endangered species." I think you go to jail for some of that. So it literally hinders what the goals are. It's the bureaucratic approach versus the free market approach. There is so much wealth in this country, there are a few billionaires around and many of them are interested in these subjects and there's no reason why with the land they own and buy and control, that they wouldn't be interested in these things.[50]

This seems to suggest that Paul would favor complete repeal of the act, even if he still supports at least some of the act's objectives. Of course, his statement reveals scientific naivete regarding the ease of propagating most endangered species; even highly motivated and well-funded private philanthropists are unlikely to be able to marshal the scientific expertise and other specialized resources needed to establish a breeding program. He might settle for an easing of the bureaucratic and economic burdens on property owners and businesses of the type favored by Romney, but Paul's opposition to the act is more than just economic—it is philosophic. He would essentially return endangered species protection to its status prior to the act. He might verbally encourage private citizens to take actions to protect species in danger of extinction, but he would remove government limitations on actions that jeopardize the continued existence of species and offer no government structure or funding to protecting species from extinction.

The fact that endangered species are not mentioned in the Constitution is not unusual; almost no commonplace federal government functions other than national defense are specifically mentioned in what is more of a national mission statement than a set of specific directions for all federal government functions. No species were perceived as being in danger of extinction at the time the Constitution was written; indeed, the very concept of extinction and the underlying science of ecology did not even exist at that time. Libertarians such as Paul commonly argue that issues such as endangered species are excluded from the purview of the federal government by the Tenth Amendment to the U.S. Constitution, which states that "the powers not delegated to the United States by the Constitution, nor prohibited by it to the States, are reserved to the States respectively, or to the people."[51]

But, such an argument is diminished when considering that the distribution of animal and plant species does not observe the political boundaries of states, or even countries, and that a well-planned program for recovering most species from near extinction requires integrated coordination among multiple states, and even countries. From a purely scientific perspective, the Endangered Species Act is therefore best suited to federal administration, and even a broader global program can be necessary for some species. This need for global action is reflected by the Convention on International Trade in Endangered Species (CITES) and the Migratory Bird Treaty Act, discussed in Chapter 1 of this book.

8.4.7.3 Rick Santorum

Rick Santorum's campaign Web site does not specifically mention the Endangered Species Act, although it speaks considerably about promoting energy development and easing regulatory burdens.[52] In a speech by Santorum in South Carolina on January 16, 2012, he stated:

> Last year, top state [South Carolina] environmental policymaker Allen Amsler wrote that the current regulatory environment is "becoming a deterrent to any business looking to move to South Carolina—or for those that are already here, expanding them." Amsler, who chairs South Carolina's Department of Health and Environmental Control board of directors, expressed his concern over the "continuous flow of new regulations" coming in from the Environmental Protection Agency, which have high costs to businesses and economic growth.[53]

It is clear from the text of this speech, as well as other information on Santorum's Web page, that he is first and foremost interested in repeal of Obama's carbon emissions regulations, especially those regulations hampering energy development, but when confronted with the Endangered Species

Act, one may reasonably expect Santorum to favor easing the effort of land-owners and private-sector businesses in complying with the act. Santorum's voting record does not suggest a proenvironment record; Republicans for Environmental Protection gave him zeroes (where zero represents a completely unfavorable environmental voting record and 100 a completely favorable environmental voting record) for both years of the 109th Congress, just prior to his leaving his Senate seat.[54]

In a speech to Colorado voters on February 6, 2012, Santorum stated:

> We have the Endangered Species Act, which has prevented us from timbering all sorts of acreage there [Pennsylvania]. It's [sic] bankrupted the school district and the like because of the government's inability to allow for us to care for our resources. A forest in my opinion is like a garden and you've got to care for it. If you don't care for it, you leave it to nature and nature will do what it does: boom and bust.[55]

Of course, natural forests of the type depended on by many endangered species are not gardens, and few endangered species benefit from conversion of natural forests to intensively managed systems. Many endangered species that depend on old-growth forest cover, such as the red-cockaded woodpecker (*Picoides borealis*), which depends on old-growth pine forests in the Southeast, and the northern spotted owl, which depends on old-growth forests of the West, require overmature trees with hollows that are generally eliminated by active forest management targeting timber harvest.

8.4.7.4 Barack Obama

Like his Republican opponents, Barack Obama does not mention the Endangered Species Act on his reelection campaign Web site,[56] and he has said little about or done little to the act over the course of his first term. Mr. Obama's environmental agenda has been highly focused on green energy and related "green jobs." Obama's White House Web site states:

> The Obama Administration is committed to protecting the air we breathe, water we drink, and land that supports and sustains us. From restoring ecosystems in the Chesapeake Bay and the Everglades, to reducing the impacts of mountaintop mining, we are bringing together Federal agencies to tackle America's greatest environmental challenges.[57]

Obama is not an outspoken environmentalist. Although there is no shortage of opinions on Obama's positions on the Endangered Species Act, my opinion is that he likely thinks little about it—perhaps no more than do any of his Republican opponents. Obama established his career primarily as a social activist, a self-described "community organizer" fighting for racial and class equality. Those objectives are in no way contradictory to those of the Endangered Species Act, but neither do they substantially complement the act's objectives. The objectives are just different—"apples and oranges" to use a common cliché. Obama's linkage to the supporters of the Endangered Species Act is through shared objectives encompassed by the Democratic Party. The League of Conservation Voters stated that "the [Endangered Species Act] was under siege for much of the George W. Bush era" and that "the Obama Administration has restored some much-needed balance when it comes to protecting endangered species."[58]

The Endangered Species Act has generally not been weakened over Obama's first term. It had generally not been weakened, at least directly, under George W. Bush's two terms either, other than for the midnight rule changes. Obama overturned the midnight rule changes early in his first year. More recently, the Obama administration proposed a new policy in December 2011 that "a plant or animal could be listed as threatened or endangered if threats occur in a 'significant portion of its range,' even if the threat crosses state lines and does not apply in the species' entire range."[59] But, in general the Endangered Species Act has not been at the center of controversy in Obama's term. Whereas wetlands and property rights issues were highly visible issues in Clinton's two terms in the 1990s, environmental issues in Obama's term have been centered mostly on climate change and renewable energy.

In December 2011, the director of the FWS proposed a policy whereby new species would be listed only if their populations were declining and in jeopardy of extinction over their entire range, not just a "significant portion of their range."[60] Critics expressed concern that this policy, if in place from the outset of the Endangered Species Act, would have prevented the listing of numerous species, including the bald eagle, which has successfully recovered over the duration of the act.[61] From the outset of the act, many species have been listed as threatened or endangered only over a portion of their range. The bald eagle is a good example; it has always remained relatively common in Alaska, and for much of its tenure listed under the Endangered Species Act, it was listed only in all or a portion of the "lower 48" states. From one perspective, the policy change has scientific merit. Extinction is the complete loss of the species; protecting species in danger of "extinction" over only a significant portion of their range is preventing "extirpation" not extinction. But, as explained in Chapter 2 of this book, the concept of extinction is quite blurry; extirpation of a species from a significant part of its range likely constitutes a significant irretrievable loss of genetic material. The new policy would therefore constitute a substantial weakening of the act. It would not merely serve to render the act more practicable.

Although this proposed policy may represent the will of the FWS director rather than President Obama per se, the director is answerable through the secretary of the interior to the president. This proposed policy illustrates that while Obama may be less inclined to weaken the Endangered Species Act than the Republican candidates, he is clearly willing to compromise on elements of the act. Such compromise, as evidenced by Clinton era compromises such as the no surprises rule, can be a good thing, making the act more palatable. But, compromises must be carefully vetted by qualified scientists before proposal by politicians.

8.5 The Future of the Endangered Species Act

At the time this book is expected to be published, the Endangered Species Act will be approaching its 40-year anniversary. NEPA has already passed its 40-year anniversary. The Clean Water Act is also 40 years old, although its most relevant aspect for endangered species, regulation of wetlands, was not firmly established until about 25 years ago. Clearly, environmental regulation by the federal government following the general pattern established in the late 1960s and early 1970s is now well established in the United States. The inspiration of the 1970s produced the Endangered Species Act and related environmental regulation, and the pragmatism of the three succeeding decades has molded these elements into the regulatory policies followed today. The Endangered Species Act and related environmental protection acts have gone through a tortuous maturation process as their idealistic objectives collided with the realities of economic and property ownership interests.

A general trend has been evident over the life of the Endangered Species Act that generally parallels that for many other environmental protection statutes. The typical lifestyle phases of an environmental statute may be postulated as follows:

1. *Agitation for the statute.* Agitation may take the form of formal protests, media campaigns, books, or general efforts to direct attention to a cause. Key elements during the agitation leading to the Endangered Species Act and other environmental statutes of the late 1960s and early 1970s include the landmark conservation works *A Sand County Almanac* by Aldo Leopold and *Silent Spring* by Rachel Carson. The emergence of public concern becomes manifested in politics, elections, and possibly, ultimately, a statute.

2. *Idealism following passage of the statute.* Supporters of the statute are relieved that the subject resource (e.g., endangered species) is now protected. Even though they may disagree on the specific regulatory practices needed to implement the new statute, they are generally pleased that the statute is in place. This is where the Endangered Species Act stood in the 1970s. The statute and its implementing regulations, once established, are usually relatively simple and in line with its founding objectives. Most of the relatively few listed

endangered and threatened species in the 1970s were well known and recognized among the American public (e.g., the bald eagle, brown pelican, and American crocodile). Although the Endangered Species Act had adversaries, public and political enthusiasm for the act overwhelmed them.

3. *Agitation against the statute.* Environmental regulatory statutes such as the Endangered Species Act do benefit the public at large, but they do not impose the costs (monetary or otherwise) equally on the public. Environmental protection statutes involving land use controls, such as the Endangered Species Act, impose costs most highly on property owners, both individuals and businesses. Combined, these two sectors form a political bloc with substantial clout; enthusiasm for voting and elections has always been highest for the most active participants in society, such as landowners and business owners. In addition, the body of regulations and sometimes even the statutory language tend to become increasingly complicated as special interests push for modifications. In the case of the Endangered Species Act, more and more environmental groups began to petition for listing of more and more species, thereby increasing the length and complexity of the list and the number of property owners affected. This was where the Endangered Species Act found itself in the early 1990s, during the pivotal 1994 elections and the Contract with America. Elements of the resulting counteragitation resemble elements of the initial proagitation phase; they also commonly involve formal protests, media campaigns, books, or general efforts to direct attention to a cause of easing the burdens of a statute.

4. *Adaptation of the statute.* This is the desired ultimate outcome. The statute, or at least its implementing regulations, is modified to make it more practicable in the real world while not dismissing the statute's objectives. Achieving this outcome may require compromise on the part of the act's supporters. The establishment of incidental take permits and the no surprises rule are examples of how the Endangered Species Act has become adapted to function better in the real world. The alternative is for a statute to be repealed or modified to the point that it no longer achieves its stated objectives. The Endangered Species Act stands at this crossroads today. It will have to continue to adapt to the complex and changing society we live in. While outright repeal of a long-standing statute such as the Endangered Species Act is unlikely, the statute can be effectively gutted even without legislative action from Congress. Agencies can repeal or change regulations, and hence regulatory policy, without any changes to the statute. Perhaps the easiest, most furtive approach to reducing the scope of a regulatory statute is for Congress simply not to fund adequately the agencies authorized to implement the statute.

Not only has this pattern been followed by the Endangered Species Act, but it also tracks reasonably well with the courses of NEPA; Section 404 of the Clean Water Act; the National Historic Preservation Act; the Comprehensive Environmental Response, Compensation, and Liability Act; and the Clean Air Act Amendments of 1990.

Environmental consultants and other practitioners working with the Endangered Species Act will almost certainly be challenged to keep up with proposed changes to the act and its regulations and policies. They have long had to follow multiple, simultaneous, and often contradictory proposals to change the regulations they deal with, as well as keep up with ongoing technical advances in biology, ecology, and other scientific fields. Assisting these practitioners will be an increased body of knowledge easily accessible through the Web or even though smart phone applications; there should be less need for slower written communication with the Services (although frequent oral communication with the Services is still highly advisable).

Most of all, environmental practitioners working with the Endangered Species Act will have to be adaptable. This is true for environmental practitioners in general. It will not be enough to be able to crank out biological assessments following a boilerplate or standardized outline. Action agencies will increasingly demand consultants with a comprehensive understanding of the act who can identify the most expeditious, not the most traditional or the most elaborate, route to successful compliance.

Notes

1. Muir, J. 1911. *My First Summer in the Sierra*, Houghton Mifflin, Boston.
2. Leopold, A. 1966. *A Sand County Almanac: With Other Essays on Conservation from Round River*, Oxford University Press, New York, 240 pp.
3. Witmer, M.C., and Cheke, A.S. 1991. The Dodo and the Tambalocoque Tree: An Obligate Mutualism Reconsidered. *Oikos* 61(1): 133–137.
4. Lanza, R.P., Cibelli, J.B., Diaz, F., Moraes, C.T., Farin, P.W., Farin, C.E., Hammer, C.J., West, M.D., and Damiani, P. 2000. Cloning of an Endangered Species (*Bos gaurus*) Using Interspecies Nuclear Transfer. *Cloning*. 2(2): 79–90.
5. Connor, S. 2009. The Big Question: Could Cloning Be the Answer to Saving Endangered Species from Extinction? *The Independent*, February 3. Available at http://www.independent.co.uk/news/science/the-big-question-could-cloning-be-the-answer-to-saving-endangered-species-from-extinction-1543657.html. Accessed March 14, 2012.
6. Psalms 104: 25, 30. Available at http://iowa.sierraclub.org/icag/2004/1104quotes.pdf. Accessed March 14, 2012.
7. Ezekiel 34: 17–18. Available at http://iowa.sierraclub.org/icag/2004/1104quotes.pdf. Accessed March 14, 2012.

8. Lev. 25: 23–24. Available at http://iowa.sierraclub.org/icag/2004/1104quotes. pdf. Accessed March 14, 2012.
9. Simmons, D.R., and Simmons, R.T. 2003. The Endangered Species Act Turns 30. Available at www.cato.org/pubs/regulation/regv26n4/mercreportcom.pdf. Accessed February 28, 2012.
10. Libertarian Party. Issue: Environment. Available at http://www.lp.org/issues/ environment. Accessed February 28, 2012.
11. U.S. Constitution, Fifth Amendment.
12. 272 U.S. 365 (1926)
13. U.S. Supreme Court Center, *Village of Euclid v. Ambler Realty Co.*, 272 U.S. 365 (1926). Available at http://supreme.justia.com/cases/federal/us/272/365/ case.html. Accessed February 26, 2012.
14. *United States v. Riverside Bayview Homes, Inc.*, 474 U.S. 121 (1985).
15. Environmental Laboratory. 1987. *Corps of Engineers Wetlands Delineation Manual,* Technical Report Y-87-l, U.S. Army Engineer Waterways Experiment Station, Vicksburg, MS.
16. Federal Interagency Committee for Wetland Delineation. 1989. *Federal Manual for the Identification and Delineation of Jurisdictional Wetlands.* U.S. Army Corps of Engineers, U.S. Environmental Protection Agency, U.S. Department of the Interior Fish and Wildlife Service, and U.S. Department of Agriculture Soil Conservation Service. Washington, DC.
17. The Soil Conservation Service is presently termed the Natural Resource Conservation Service.
18. 16 U.S.C. 3801–3862.
19. Speech by Vice President Dan Quayle to Competitive Enterprise Institute, Washington, DC, April 21, 1999; as cited in On the Issues, Every Political Leader on Every Issue: Dan Quayle on Environment. Available at http://www.issues2000. org/Celeb/Dan_Quayle_Environment.htm. Accessed February 26, 2012.
20. Ibid.
21. U.S. Fish and Wildlife Service. 2011. Revised Recovery Plan for the Northern Spotted Owl (*Strix occidentalis caurina*). U.S. Fish and Wildlife Service, Portland, OR, 258 pp. plus appendices, Appendix A.
22. Ibid., Appendix B.
23. Ibid., Appendix A.
24. 54 FR 26666.
25. 55 FR 26114–26194, June 26, 1990, p. 26124.
26. U.S. Fish and Wildlife Service. 2002. Recovery Plan for the California Red-legged Frog (*Rana aurora draytonii*). U.S. Fish and Wildlife Service, Portland, OR, 173 pp. plus appendices, Executive Summary.
27. Mark Twain. 1865. The Celebrated Jumping Frog of Calaveras County. Available at http://etext.virginia.edu/railton/huckfinn/jumpfrog.html. Accessed February 27, 2012.
28. Republican Contract with America. Available at http://www.house.gov/ house/Contract/CONTRACT.html. Accessed February 25, 2012.
29. Summary of the Job Creation and Wage Enhancement Act. Available at http:// www.house.gov/house/Contract/cre8jobsd.txt. Accessed February 25, 2012.
30. 104th Congress, 1st Session, H.R. 925, In the Senate of the United States, March 7 (legislative day, March 6), 1995. Available at http://thomas.loc.gov/cgi-bin/ query/D?c104:4:./temp/~c104ebP8Qa. Accessed February 25, 2012.

31. H.R.925—Private Property Protection Act of 1995 (Reported in House, RH). Available at http://thomas.loc.gov/cgi-bin/query/D?c104:1:./temp/~c104US-Jmae. Accessed February 25, 2012.
32. The Supreme Court consists of nine justices who vote on decisions; hence, a 5-4 decision indicates a divisive "close call" with substantial support both for and against the decision.
33. 474 U.S. 121 (1985).
34. 51 FR 41217 (1986), Migratory Bird Rule.
35. 531 U.S. 159 (2001).
36. Opinion of Chief Justice Rehnquist regarding *SWANCC v. Army Corps of Engineers*, January 9, 2001.
37. Home page, http://www.sws.org.
38. 547 U.S. 715 (2006).
39. Strictly speaking, the Rapanos forms were to be completed by USACE, but USACE strongly encouraged Section 404 permit applicants (or their consultants) to complete the forms to expedite application processing.
40. H.R. 3824—109th Congress: Threatened and Endangered Species Recovery Act of 2005. 2005. In GovTrack.us (database of federal legislation). Available at http://www.govtrack.us/congress/bill.xpd?bill=h109-3824&tab=summary. Accessed February 25, 2012.
41. 73 FR 28212–28303, May 15, 2008.
42. Greenemeier, L. 2008. U.S. Protects Polar Bears Under Endangered Species Act. May 14. Available at http://www.scientificamerican.com/article.cfm?id=polar-bears-threatened&page=2. Accessed March 10, 2012.
43. 75 FR 76086–76137, December 7, 2010.
44. 77 FR 4492–4493, January 30, 2012.
45. Rep. John Dingell. 2009. Dingell Praises Obama Decision to Reverse Bush Changes to the Endangered Species Act. Press release dated March 9. Available at http://dingell.house.gov/news/press-releases/2009/03/090303ESA.shtml. Accessed March 11, 2012.
46. World Wildlife Fund. 2008. Midnight Rule Changes by Bush Administration Will Undermine Endangered Species Protections, Says WWF, Polar Bear Listing Weakened by Administration's "Parting Shots." December 12. Available at http://www.worldwildlife.org/who/media/press/2008/WWFPresitem 11028.html. Accessed March 10, 2012.
47. 40 C.F.R. 1500–1508.
48. Romney for President. Regulation. Available at http://www.mittromney.com/issues/regulation. Accessed February 26, 2012.
49. Ron Paul for President. Energy. Available at http://www.ronpaul2012.com/the-issues/energy/. Accessed February 26, 2012.
50. Postman, D. 2007. Ron Paul Says Rich Should Fund Species Protection. *The Seattle Times*, September 7. http://blog.seattletimes.nwsource.com/davidpostman/archives/2007/09/ron_paul_says_rich_should_fund_species_protection.html. Accessed February 28, 2012.
51. U.S. Constitution, Tenth Amendment.
52. Rick Santorum 2012 Presidential Campaign. Issues. Available at http://www.ricksantorum.com/issues. Accessed March 8, 2012.

53. Santorum, R. 2012. Stop Crushing Economy with Regulations. January 16. Available at http://www.ricksantorum.com/obama-regs-versus-freedom. Accessed Marsh 8, 2012.
54. Republicans for Environmental Protection. 2006. Congressional Scorecard for 2006. Available at http://www.rep.org/2006_scorecard.pdf. Accessed March 11, 2012.
55. Hooper, T. 2012. Santorum and Gingrich Dismiss Climate Change, Vow to Dismantle the EPA. *The Colorado Independent*. February 6. Available at http://coloradoindependent.com/111924/santorum-and-gingrich-dismiss-climate-change-vow-to-dismantle-the-epa. Accessed March 8, 2012.
56. Obama–Biden 2012 Reelection. The President's Record on Energy and the Environment. Available at http://www.barackobama.com/record/environment?source=footer-nav. Accessed March 11, 2012.
57. The White House. Our Environment. Available at http://www.whitehouse.gov/energy/our-environment. Accessed March 11, 2012.
58. League of Conservation Voters. 2012. Issues: Wildlife. Available at http://www.lcv.org/issues/wildlife/. Accessed March 11, 2012.
59. Fox News. 2011. Admin[istration] Moves to Clarify Endangered Species Listings. Available at http://www.foxnews.com/us/2011/12/09/apnewsbreak-no-more-species-listings-by-state/. Accessed March 11, 2012.
60. 76 FR 76987, December 9, 2011.
61. Letter dated January 26, 2012, from Rep. Edward J. Markey of Massachusetts to Dan Ashe, Director, U.S. Fish and Wildlife Service. Available at http://www.biologicaldiversity.org/programs/biodiversity/endangered_species_act/pdfs/Markey_Letter.pdf. Accessed March 12, 2012.

Index

Note: **CI** indicates color insert

9 781138 374676